Discrete Element Method for Multiphase Flows with Biogenic Particles

Ling Zhou • Mahmoud A. Elemam
Ramesh K. Agarwal • Weidong Shi

Discrete Element Method for Multiphase Flows with Biogenic Particles

Agriculture Applications

Springer

Ling Zhou
Research Center of Fluid Machinery
Engineering and Technology
Jiangsu University
Zhenjiang, Jiangsu, China

Mahmoud A. Elemam
Research Center of Fluid Machinery
Engineering and Technology
Jiangsu University
Zhenjiang, Jiangsu, China

Department of Agricultural and
Biosystems Engineering
Alexandria University
Alexandria, Egypt

Ramesh K. Agarwal
Department of Mechanical Engineering
and Materials Science
Washington University in Saint Louis
Saint Louis, MO, USA

Weidong Shi
School of Mechanical Engineering
Jiangsu University
Nantong, Jiangsu, China

ISBN 978-3-031-67728-1 ISBN 978-3-031-67729-8 (eBook)
https://doi.org/10.1007/978-3-031-67729-8

This Springer imprint is published by the registered company Springer Nature Switzerland AG
The registered company address is: Gewerbestrasse 11, 6330 Cham, Switzerland

Preface

In recent years, within the realm of aerodynamic systems involving biogenic particulate matter, the complexities of particle motion in the fluid flow have posed unique challenges, necessitating innovative modeling approaches. This book provides a comprehensive review of recent strategies and applications centered around the numerical simulations coupling the Computational Fluid Dynamics methods with the Discrete Element Method (CFD–DEM). The primary focus of the book is on the biogenic particulate matter such as grains, crops, and biomass, exploring the limitations of the traditional continuum-based modeling and advocating for the promising capabilities of CFD–DEM modeling approach in capturing the discontinuous behavior of these particles. The narrative unfolds by thoroughly examining the aerodynamic systems establishing a foundational understanding of the study's principles. The subsequent exploration into the modeling of the particle motion reveals the inadequacies of the conventional approaches, setting the stage for introducing the CFD–DEM coupling methodology. This innovative numerical method, which combines the strengths of Computational Fluid Dynamics and the Discrete Element Method, emerges as a powerful tool for tracking the individual trajectories of the particles in fluid flow. The agricultural field provides a pertinent backdrop for the applications discussed in this book. Focusing on the grains, crops, and biomass, the book sheds light on the potential benefits of employing the CFD–DEM approach to unravel the complex interplay between the bio-particulate matter and fluid dynamics. The coupling strategy is shown to be particularly promising, offering insights that traditional methods struggle to provide and laying the groundwork for optimizing and enhancing the various processes within the agricultural domain. The findings presented in this book culminate in a resounding endorsement of CFD–DEM as a valuable approach for scrutinizing the behavior of the biogenic particulate matter in fluid flow. From the drag correlation models to the specific applications in agriculture, the book describes the breadth of possibilities that this coupling methodology provides. The last chapter on concluding remarks emphasizes the promise inherent in the CFD–DEM method and spotlights the future developments and challenges awaiting in exploration of this dynamic field. By synthesizing cutting-edge research, practical applications, and prospects, this book aims to be a cornerstone for scholars, researchers, and practitioners seeking to deepen their understanding of the CFD–DEM coupling approach in the context of aerodynamic systems involving biogenic particulate matter. As readers navigate through the pages of the book, the intricacies of the granular flow and fluid dynamics unfold, inviting them to embark on a journey of discovery and innovation at the intersection of fluid mechanics and computational modeling with application to problems encountered in agriculture.

Zhenjiang, Jiangsu, China — Ling Zhou
Zhenjiang, Jiangsu, China — Mahmoud A. Elemam
Saint Louis, MO, USA — Ramesh K. Agarwal
Nantong, Jiangsu, China — Weidong Shi

Acknowledgments

The publication of the book is not supported by these grants. The research was supported. The research presented in this book was supported by the National Natural Science Foundation of China (Grant No. 52079058), Natural Science Foundation of Jiangsu Province (Grant No. BK20230011), and Jiangsu Funding Program for Excellent Postdoctoral Talent (Grant No. 2022ZB641). The resources provided by the Jiangsu University, Zhenjiang, China, and Washington University in St. Louis, MO, USA, are gratefully acknowledged. Finally, the authors express their gratitude to their respective family members for their encouragement and unwavering support.

Contents

Nomenclature

ϕ_{ij}	pressure strain
$\phi_{ij,1}$	slow pressure-strain
$\phi_{ij,2}$	rapid pressure-strain
$\phi_{ij,3}$	wall-reflection term
η	damping ratio
η^t	tangential damping ratio
ρ_a	air density
ρ_f	fluid density
$\rho_f \overline{u_i' u_j'}$	Reynolds stress tensor
ρ_i	particle density
μ	friction coefficient
μ_a	dynamic viscosity of the air
μ_f	fluid turbulent viscosity
$\mu_{f_{eff}}$	effective turbulent viscosity of a fluid
μ_t	turbulent viscosity
α_f	fluid volume fraction
α_i	the particle volume fraction
α_k & α_ε	inverse effective Prandtl number for k and ϵ
σ_k & σ_ε	turbulent Prandtl numbers for k and ϵ
ε	coefficient of restitution
ε_{ij}	dissipation
δ_{adh}	adhesive distance
τ_f	viscous stress tensor
υ_f	kinematic viscosity of the fluid
ΔS^n	change in the normal overlapping
ΔS^t	change in the tangential overlapping
ΔT	timestep
Γ	surface energy
$\varnothing_i$	particle sphericity
a	separation radius between particles
A_i	the projected area of particle

C_d	drag coefficient
C_{ij}	convection
d	normal distance to the wall
$\frac{d}{dT}S^t$	relative tangential velocity at the contact
d_i	particle diameter
$d_{i_{vol}}$	particle diameter with the same volume as an actual one
D	domain diameter
$D_{L,ij}$	molecular diffusion
$D_{T,ij}$	turbulent diffusion
E_i	Young's modulus of particle
E_w	Young's modulus of wall boundary
f_{adh}	adhesive force fraction
F_{ij}	force production by system rotation
F_{net}	the net force acting on a particle
F_{ij}^C	total contact force acting on the particle
F_i^f	fluid-particle interaction force
f_i^g	gravitational force
F_{ij}^n	normal contact force on the particle
$F_i^{n,T}$	normal contact force at the current time
$F_i^{n,(T-\Delta T)}$	normal contact force at a previous time
$F_{i_{adh}}^{n,T}$	normal adhesive force at the current time
F_{ij}^t	tangential contact force on the particle
$F_i^{t,T}$	tangential contact force at the current time
$F_i^{t,(T-\Delta T)}$	tangential contact force at the previous time
g	gravity acceleration
G_k	kinetic energy generation due to the velocity
G_b	kinetic energy generation due to buoyancy
G_{ij}	the buoyancy production rate of $\overline{u_i' u_j'}$
K_1	Stokes' shape factor
K_2	Newton's shape factor
k_a	coefficient of aerodynamic resistance
K_l^n	loading contact stiffnesses
K_u^n	unloading contact stiffnesses
l	the length scale of energy-containing eddies
m^*	equivalent mass of particles
m_i & m_j	mass of particles i & j
N	number of particles
$\widehat{n}_c$	the unit vector in the normal direction
n_k	component of x_k normal to the wall
n & t	normal and tangential coordinates
p	pressure shared by two phases
P_{ij}	the stress production rate of $\overline{u_i' u_j'}$
P_{max}	maximum liquid bridge tensile force
R_ε	dissipation of swirl and rotational effect

R_e	Reynolds number
R_{e_i}	particle Reynolds number
R_i & R_j	radii of particles i & j
S_k & S_ε	constant source terms for user-defined
S^n	normal overlapping
$S^{n,T}$	normal overlap value at the current time
$S^{n,T-\Delta T}$	normal overlap value at the previous time
s^t	tangential relative displacement at the contact
s^t_{max}	maximum relative tangential displacement
S_{user}	constant source term for user-defined
T	time
T_e	turnover time of large eddy
T_i	response time of a particle
T_K	Kolmogorov time scale
u	root mean square of fluid velocity
u_f	average velocity component of a fluid
V_a	air velocity
V_f	the average velocity of the fluid
V_i & V_j	particles velocity before interaction
V_i' & V_j'	particles velocity after interaction
V_T	terminal velocity
Vol	the volume occupied by particles and fluid
Vol_i	particle volume
Y_M	dilatation contribution over dissipation rate
Z	particle size or geometry

Chapter 1
Introduction

The modeling of multiphase flows in aerodynamic systems involving biogenic particles constitutes a complex and interdisciplinary field that seamlessly integrates computational fluid dynamics (CFD) with the discrete element method (DEM). The coupling between CFD and DEM enables the simulation of the intricate behaviors of multiphase flows featuring biogenic particles [1, 2]. This coupled CFD–DEM approach for numerical simulations holds a great deal of promise across diverse scientific and engineering fields, spanning environmental science to industrial processes. At its core lies the intricate interplay between fluid dynamics and particulate matter—a phenomenon of great complexity [3, 4]. A thorough understanding of this interaction is indispensable for optimizing systems, enhancing efficiency, and minimizing environmental impact. Biogenic particles, organic particles sourced from living organisms, are of particular significance within this domain. These particles manifest in various natural processes, such as atmospheric aerosols, soil erosion, and biological fluid dynamics [5–7]. The modeling of the aerodynamic characteristics of these particles in multiphase flows yields valuable insights into dispersion, transport, and deposition mechanisms. These insights are crucial for predicting the environmental repercussions and fine-tuning industrial processes for optimal performance.

The preparation and processing of biogenic raw materials and various grains, including herbs, involve a spectrum of technological procedures. These processes often necessitate specific steps, such as purification and separation into distinct-size fractions [8–10]. In the cereal and fodder industries, the separation of mixtures containing biogenic particles and other loose materials of biological origin stands out as a pivotal technological operation [11, 12]. The efficiency of these procedures significantly influences subsequent processes and the ultimate quality of the end product, whether it be seeding material, flour purity, groats, cereal flakes, and the like [13]. The separation of grain mixtures hinges on leveraging the differences in physical properties between the general material targeted for purification and

L. Zhou et al., *Discrete Element Method for Multiphase Flows with Biogenic Particles*, https://doi.org/10.1007/978-3-031-67729-8_1

separation and the constituents of the mixture, which are typically contaminants [14–17]. The success of this separation process is instrumental in determining the quality of the final product.

The significance of CFD–DEM modeling in this context is rooted in its capacity to intricately capture the interactions between fluids and particles at a granular level. In contrast to conventional approaches that may oversimplify particle behavior in a fluid medium, CFD–DEM modeling enables a more realistic representation of the intricate physics at play [18]. This becomes especially critical when dealing with biogenic particles, where their organic nature introduces additional complexities that demand accurate consideration in simulations. The implications of such simulations are far-reaching [19, 20]. In environmental science, for instance, comprehending the aerodynamic systems of multiphase flows involving biogenic particles facilitates predicting pollutant dispersion, assessing air quality, and formulating strategies to mitigate the impact of natural phenomena like dust storms [21]. In industrial applications, optimizing processes related to biogenic particles, such as pharmaceutical production or handling agricultural materials, can result in heightened efficiency, reduced waste, and improved overall sustainability. Achieving these objectives necessitates advanced numerical techniques, high-performance computing, and sophisticated algorithms [22, 23]. The CFD component involves numerical fluid flow simulation using governing equations that describe the conservation of mass, momentum, and energy. Conversely, DEM is dedicated to modeling the discrete behavior of particles, accounting for their interactions with each other and the surrounding fluid.

The integration of CFD and DEM involves the solution of coupled equations, intricately linking the fluid flow and the particle dynamics. This coupling enables a more precise representation of real-world interactions, encompassing particle-particle collisions, fluid-particle interactions, and the impact of turbulence on particle dispersion [24–26]. The technological solutions in this aspect include the development of specialized software and simulation platforms tailored to tackle the challenges posed by multiphase flows involving biogenic particles. In conclusion, the modeling of aerodynamic systems with multiphase flows containing biogenic particles is an advanced and important field of research with broad applications in science and engineering. Its significance spans environmental impact assessments to optimization of industrial processes, illustrating the need for accurate and realistic numerical simulations. The technology solutions described in this book include the continuous advancements in numerical methods, computational tools, and simulation platforms that contribute to a more complete understanding of the intricate interactions between fluid dynamics and biogenic particles. As research in this field advances, it has the potential to drive innovation, inform sustainable practices, and address the challenges posed by multiphase systems involving biogenic particles. It is essential to emphasize that the separation processes in the agricultural and food industries are notably intricate compared to other branches of mechanics involving granular flows. This complexity arises from the biologically diverse nature of the materials involved, characterized by significant variability and low reproducibility even within the same grain variety or species. The interdisciplinary nature of the physical traits of the constituents within a mixture, coupled with the varying technical parameters of pneumatic separation processes, motivates the exploration of the subject matter presented in this book.

References

1. Scherer V, Mönnigmann M, Berner MO, Sudbrock F (2016) Coupled DEM–CFD simulation of drying wood chips in a rotary drum—baffle design and model reduction. Fuel 184:896–904. https://doi.org/10.1016/j.fuel.2016.05.054
2. El-Emam MA, Zhou L, Shi W et al (2021) Theories and applications of CFD—DEM coupling approach for granular flow: a review. Arch Comput Methods Eng 28:4979–5020. https://doi.org/10.1007/s11831-021-09568-9
3. Zhao Z, Zhou L, Bai L et al (2023) Modeling and validation of coarse-grained computational fluid dynamics-discrete element method for dense gas-solid flow simulation in a bubbling fluidized bed. Phys Fluids 35:043310. https://doi.org/10.1063/5.0146264
4. Fonte CB, Oliveira JAA, De Almeida LC (2015) Dem-Cfd coupling: mathematical modelling and case studies using Rocky-Dem® and Ansys Fluent®. In: Eleventh international conference on CFD in the minerals and process industries, Organ, pp CSIRO, 1–CSIR7
5. Ma H, Zhou L, Liu Z et al (2022) A review of recent development for the CFD-DEM investigations of non-spherical particles. Powder Technol 412:117972. https://doi.org/10.1016/j.powtec.2022.117972
6. Chemical T (1988) Effectiveness factors for bioparticles with Monod kinetics. Chem Eng J 37:31–37
7. Li J, Webb C, Pandiella SS, Campbell GM (2002) A numerical simulation of separation of crop seeds by screening – effect of particle bed depth. Food Bioprod Process Trans Inst Chem Eng Part C 80:109–117. https://doi.org/10.1205/09603080252938744
8. Ma Z, Li Y, Xu L (2015) Discrete-element method simulation of agricultural particles' motion in variable-amplitude screen box. Comput Electron Agric 118:92–99. https://doi.org/10.1016/j.compag.2015.08.030
9. Li J, Webb C, Pandiella SS, Campbell GM (2003) Discrete particle motion on sieves – a numerical study using the DEM simulation. Powder Technol 133:190–202. https://doi.org/10.1016/S0032-5910(03)00092-5
10. Lei X, Liao Y, Liao Q (2016) Simulation of seed motion in seed feeding device with DEM-CFD coupling approach for rapeseed and wheat. Comput Electron Agric 131:29–39. https://doi.org/10.1016/j.compag.2016.11.006
11. Williams GD, Rosentrater KA (2022) Design and engineering considerations for grain storage, handling, and processing facilities. In: Rosentrater KA (ed) Storage of cereal grains and their products, 5th edn. Woodhead Publishing, Cambridge, pp 179–246
12. Yuan J, Wu C, Li H et al (2018) Movement rules and screening characteristics of rice-threshed mixture separation through a cylinder sieve. Comput Electron Agric 154:320–329. https://doi.org/10.1016/j.compag.2018.09.012
13. Bracacescu C, Gageanu I, Popescu S, Selvi KC (2016) Researches concerning impurities separation process from mass of cereal seeds using vibrating sieves in air flow currents. In: International scientific conference, pp. 364–370
14. Bettge A, Rubenthaler GL, Pomeranz Y (1989) Air-aspirated wheat cleaning in grading and in separation by functional properties. Am Assoc Cereal Chem 66:15–18
15. El-Emam MA, Zhou L, Shi WD, Sobhi M (2023) Determination of some engineering properties of jojoba granular matter [Simmondsia chinensis]. Granul Matter 25. https://doi.org/10.1007/s10035-023-01313-2

16. Zhang C, Zhuang K, Chen L et al (2022) Reducing deoxynivalenol content in wheat by a combination of gravity separation and milling and characterization of the flours produced. J Cereal Sci 104:103372. https://doi.org/10.1016/j.jcs.2021.103372
17. Güner M (2007) Pneumatic conveying characteristics of some agricultural seeds. J Food Eng 80:904–913. https://doi.org/10.1016/j.jfoodeng.2006.08.010
18. Zhang R, Ku X, Yang S (2023) Study of air-steam gasification of spherocylindrical biomass particles in a fluidized bed by using CFD-DEM coupling with the multi-sphere model. Energy Convers Manag 276:116561. https://doi.org/10.1016/j.enconman.2022.116561
19. Guo Y, Curtis JS (2015) Discrete element method simulations for complex granular flows. Annu Rev Fluid Mech 47:21–46. https://doi.org/10.1146/annurev-fluid-010814-014644
20. Mousaviraad M, Tekeste MZ, Rosentrater KA (2017) Calibration and validation of a discrete element model of corn using grain flow simulation in a commercial screw grain auger. Trans ASABE 60:1403–1415. https://doi.org/10.13031/trans.12200
21. Sakura GB, Leung AYT (2015) Experimental study of particle collection efficiency of cylindrical inlet type cyclone separator. Int J Environ Sci Dev 6:160–164. https://doi.org/10.7763/ijesd.2015.v6.581
22. Violeau D, Issa R (2007) Numerical modelling of complex turbulent free-surface flows with the SPH method: an overview. Int J Numer Methods Fluids 53:277–304. https://doi.org/10.1002/fld.1292
23. Zhou JW, Liu Y, Liu SY et al (2017) Effects of particle shape and swirling intensity on elbow erosion in dilute-phase pneumatic conveying. Wear 380–381:66–77. https://doi.org/10.1016/j.wear.2017.03.009
24. Sturm M, Wirtz S, Scherer V, Denecke J (2010) Coupled DEM-CFD simulation of pneumatically conveyed granular media. Chem Eng Technol 33:1184–1192. https://doi.org/10.1002/ceat.201000162
25. Zhao J, Shan T (2013) Coupled CFD-DEM simulation of fluid-particle interaction in geome chanics. Powder Technol 239:248–258. https://doi.org/10.1016/j.powtec.2013.02.003
26. Wang Z, Teng Y, Liu M (2019) A semi-resolved CFD–DEM approach for particulate flows with kernel based approximation and Hilbert curve based searching strategy. J Comput Phys 384:151–169. https://doi.org/10.1016/j.jcp.2019.01.017

Chapter 2
Aerodynamic Systems

Abstract For environmental science and agricultural engineering, the study of aerodynamic separation systems for biogenic particles holds significant promise and interest. This chapter describes the innovative methods and technologies designed to efficiently separate and capture biogenic particles from various sources, such as agricultural activities, biomass combustion, and other natural processes. The vertical, horizontal, inclined, and centrifugal systems present unique advantages and applications within the field. The vertical setups provide precise control; horizontal designs offer versatility; inclined devices serve some specialized purposes; and centrifugal separators boost high-throughput capabilities. In this chapter, the readers will gain a comprehensive understanding of the principles of these systems, their functionalities, and recent advancements driving their development. This chapter aims to shed light on the current landscape, challenges, and future directions of aerodynamic separation systems for biogenic particles and facilitate their use in exploring the mechanical behavior of granular flows. Emphasizing enhanced efficiency, reduced energy consumption, and improved particle purity, this chapter should serve as a valuable resource for researchers, engineers, and practitioners looking to employ and leverage aerodynamic separation systems across various industries and scientific pursuits.

Keywords Separation efficiency · Separation systems · Granular matter · Aerodynamic characteristics

2.1 Aerodynamic Systems: Classifications and Technology Status

Biogenetic granular matter has long relied on aerodynamic devices to separate, handle, and transport seeds and grains. These devices utilize air flow generated by natural or mechanical fans to effectively remove chaff, dust, and other lightweight particles from the grain. This technique is considered an essential step in the process of separating and cleaning heterogeneous mixtures of agricultural grains and can be used alone or in combination with other methods [1–3]. The benefits of this approach

L. Zhou et al., *Discrete Element Method for Multiphase Flows with Biogenic Particles*, https://doi.org/10.1007/978-3-031-67729-8_2

are manifold. It enables the simultaneous cleaning and segregation of seed particles without altering separation parameters, resulting in an undamaged product that retains its physical and biological properties. Moreover, this process is highly efficient, requiring minimal power while maintaining high precision [4]. The quality of the aerodynamic system depends on several factors, including the aerodynamic properties of the particulate matter, feeding load ratio of granular flow, speed and angle of feed grain material, airflow properties, particle properties, and the system inlet geometry [5–8].

The key to this process lies in the aerodynamic properties of the particles, such as their critical or terminal velocity and drag coefficient, which are the primary determinants of their behavior in the system [3, 9]. The greater the difference in terminal velocities between the components in the mixture, the more influential the aerodynamic separation will be [10, 11]. Aerodynamic systems can be broadly categorized based on the direction of the air stream in the system, such as vertical, horizontal, or oblique, as well as the airflow conditions or the location of the fan in the system, such as vacuum, overpressure, or hybrid systems [11, 12]. In evaluating the quality of the aerodynamic separation process, it is crucial to consider the redistribution of product components between fractions, mainly when the initial product is divided into three or more fractions [13]. With continued research and development, this technique holds great promise for enhancing the efficiency and precision of agricultural production.

2.2 Vertical Aerodynamic System

Agricultural particulate matter mixtures can be separated using a vertical air stream, but only if there is a significant difference in the aerodynamic properties of the particles. This method is effective when heavier particles fall through the air stream while lighter particles are carried along with it, but it becomes inefficient when the particles have similar characteristics [14]. Conversely, separating grain and straw mixtures is more complex, mainly when the straw pieces are short [15, 16]. Previous studies have examined the flow of grain and chaff in a vertical air stream column [2]. In this technique, the grains fall downward against the air stream while the air transports the chaff and moves upward depending on the critical velocities of each particle. The key to achieving optimal separation is selecting the appropriate air velocity that balances grain loss and undesired materials content. This can be achieved by injecting the particles into a specific region of the air column, carefully selecting the injection speed and direction relative to the air velocity, improving the way grains are fed into the system, optimizing the natural composition of the mixture, and improving the airflow within the system [8, 17].

Uhl and Lamp [2] delved into the intricacies of a separation process for soybeans, corn, oats, wheat, and rye, focusing on achieving the isolation of each component from foreign materials using a specially designed vertical air stream with specific air velocities. The researchers discovered that by reducing the straw length to 50 mm and below, complete separation of wheat and rye was achievable, while a straw length of 150 mm and below was sufficient for the complete separation of foreign materials from soybean grain. The study further revealed that rye, wheat, and oat straw could be separated with minimal grain loss, with approximately 98%, 94%, and 80% separation efficiency, respectively, while chaff-like materials could be wholly separated without grain loss. However, separating corn, cobs, and stover without grain loss was impossible. Interestingly, Gorial and O'Callaghan [18] observed that pieces of straw with a node at one end aligned vertically in the air stream, giving them a higher terminal velocity than node-less or middle-node straw, took up a horizontal position in the vertical air stream. They also developed a mathematical model in another study [15] to describe the separation process of wheat straw. They predicted the particle pathlines and the effect of injection direction in the vertical air stream. The results showed that the separation process significantly depended on the airstream direction, while the injection speed and direction had a significant effect on the air column cross-section design and dimensions. Tylek and Walczyk [19] established a theoretical basis for the aerodynamic flow of Norway spruce seeds and conducted laboratory tests to evaluate the effectiveness of the separation process in the vertical air column. Their separator was designed to separate the grains into three fractions based on critical velocities and density differences. The results demonstrated that determining the seed feeding rate and the critical velocity in the aerodynamic system significantly impacted the separation effectiveness. In addition to crops, Arinze et al. and Gan-Mor et al. [20, 21] presented a schematic frame of a vertical aerodynamic system, as shown in Fig. 2.1, for separating leaves and stems of forages, herbs, and spices, among others. The study found that the separation efficiency decreased due to particle collision with domain walls,

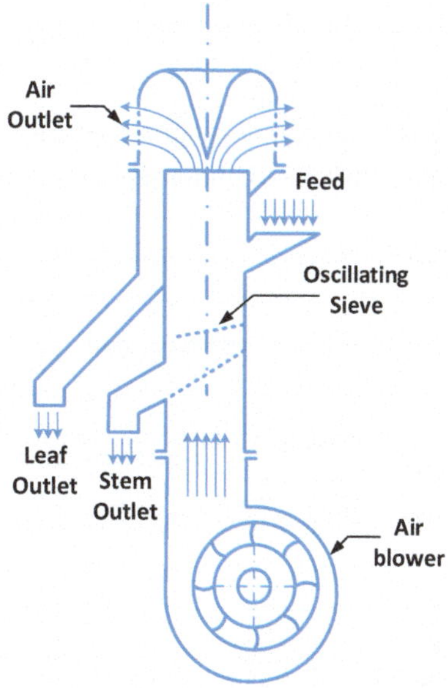

Fig. 2.1 Schematic line drawing of a vertical aerodynamic system [46]

low air velocity near the walls, and high particle feed rates, leading to particle clustering within the air stream. However, using an oscillating sieve device in the air stream helped solve this problem; it prevented the leaves from moving downward, resulting in periodic changes in particle orientation and aerodynamic drag force.

In conclusion, separating agricultural particulate matter mixtures using a vertical air stream is a complex but effective method, provided that the particles' aerodynamic properties are significantly different from each other. Separating grain and straw mixtures is a more challenging task but can be accomplished by carefully considering several factors, including injection methods, particle velocities, and airflow optimization. Using these techniques makes it possible to achieve a high level of separation effectiveness without compromising the quality of the final product. These studies have contributed significantly toward understanding the complexities of separation processes in various agricultural applications, offering valuable insights for optimizing and improving these systems.

2.3 Horizontal Aerodynamic System

In the realm of air-flow mechanics, the horizontal movement of air and particles is intricately tied to the suspension velocity of the particles. Research conducted by Gorial and O'Callaghan [22] has indicated that separating mixtures into various fractions is significantly more straightforward to accomplish with a horizontal air stream than a vertical one. To put this theory to the test, they implemented a pushed-air horizontal system, as shown in Fig. 2.2a, and utilized a diverse range of mixtures, including wheat mixed with non-grain materials, wheat and oilseed rape mixtures, soybean and grass mixtures, and a variety of other grains including chickpeas, adzuki beans, mung, sorghum, millet, and oilseed rape. Under carefully controlled conditions, they achieved a remarkable 95% separation of grain from weed at an air velocity of 11 m/s. However, it is worth noting that grading aerodynamically challenging small-sized grains, such as oilseed rape and millet, proved to be more difficult. Ultimately, these results indicate the delicate balance between a

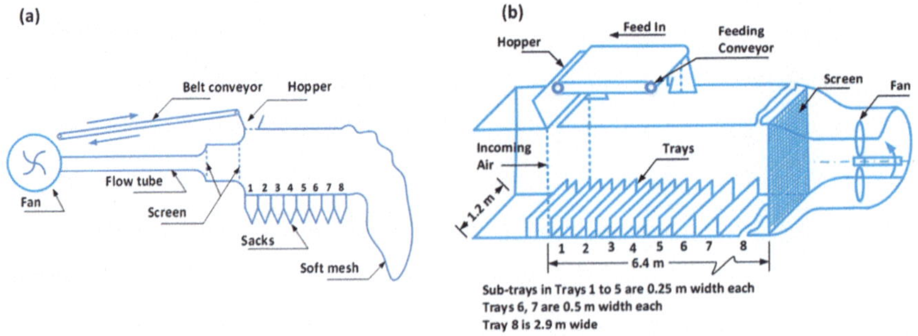

Fig. 2.2 Airflow in horizontal aerodynamic systems: (**a**) pushed air [22], (**b**) suction air [23]

particle's tendency to settle under gravity and its tendency to drift horizontally under the drag forces of an air stream.

To improve the separation of alfalfa silage into stem and leaf fractions, Bilanski et al. [23] conducted a groundbreaking experiment. Using a suction-air horizontal tunnel, as shown in Fig. 2.2b, they measured the moisture content of the silage to ensure precise separation. The aerodynamic drag coefficient, feed height level, and terminal velocity are differentiating characteristics during the separation process because alfalfa leaves and stems are remarkably similar in size and density. Because alfalfa leaves have a drag coefficient higher than that of stems [24, 25], the aerodynamic fraction would seem possible. The result showed that the best separation occurred by increasing the airstream height beyond 1 m, decreasing the feed rate and moisture content, and increasing the air velocity to 8 ms^{-1}. At this higher velocity, many fine leaf particles can be lost.

Simonyan and Yiljep [26] focused on rasp-bar sorghum's separation and cleaning efficiency using a stationary grain thresher that coupled the sieve and horizontal air stream. The results showed that the cleaning efficiency of grains decreased with the increasing feed rate, while the losses increased with increasing airspeed. These findings align with data from other research [13, 27], which highlights the importance of determining the optimum values of the feed and flow rates to optimize the grain separation process in horizontal aerodynamic systems. Overall, these studies shed light on the importance of considering various parameters when attempting to separate different fractions of crops using aerodynamic systems. This knowledge could prove valuable in enhancing the efficiency and effectiveness of crop processing techniques in the agricultural industry.

2.4 Inclined Aerodynamic System

The oblique aerodynamic system is often used to separate and sort straw, grain, and chaff mixtures in combined harvesters [28]. The efficacy of this process relies heavily on two factors: the inclination angle and the velocity of the airflow over the sieve. Studies have shown that adjusting the airstream direction to fall between 40° and 45° and maintaining an air velocity of 9.2 m/s results in maximum separation efficiency and minimal grain loss [29]. Innovative approaches to the oblique stream method have been explored by researchers, such as Panasiewicz et al. [11], who devised a system for separating and cleaning crushed lupine seeds, as shown in Fig. 2.3. The tests were conducted at various airflow velocities of 7.8, 10.5, and 12.8 m/s in combination with different moisture levels of the seeds. The results indicated that higher airflow velocities improved separation effectiveness and decreased the amount of valuable fraction carried away by the airflow. However, an increase in the moisture content caused a significant reduction in the separation effectiveness index. The success of the oblique aerodynamic system is highly dependent on the optimization of several critical parameters, including the angle of inclination and the velocity of the airflow. By continuing to explore and refine

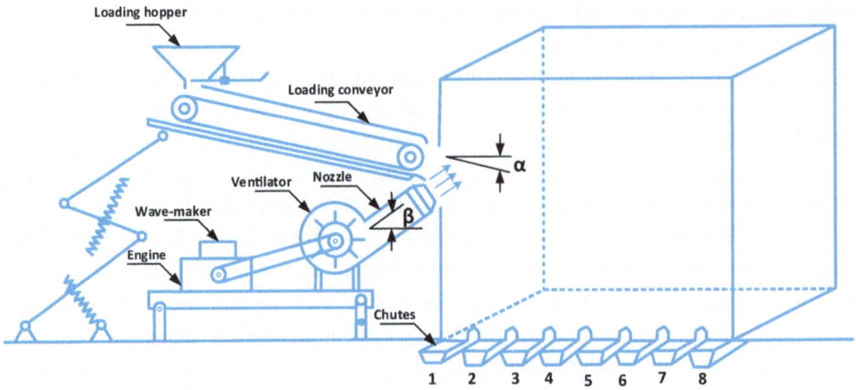

Fig. 2.3 Schematic of the oblique air stream entering the inclined aerodynamic system [11]

innovative approaches to this technique, we can further improve agricultural machinery's efficacy and enhance crop harvesting's overall efficiency.

2.5 Centrifugal Aerodynamic System

The efficient separation of biological samples with varying particle sizes and shapes has been a long-standing challenge in industries involved in simulation, modeling, and processing [30]. In the past, cyclone separators have been the go-to solution for separating multiple components of suspended particles from a gas stream in the bio-industry and bio-systems [31]. Having been in use since the nineteenth century, these devices harness centrifugal force generated by the swirling motion of fluid to separate the particles. Even today, gas cyclone separators remain among the most widely used devices in the gas-cleaning industry. Their simple design, low manufacturing and maintenance costs, ability to operate over a wide range of loading particles, and energy efficiency make them a preferred choice [32–35]. With a high separation efficiency that can reach 99% for particles larger than 5 μm, the cyclone is influenced by various factors, particularly the inlet flow velocity and particle mass loading ratio [36].

Previous research has divided aero-type cyclones into two primary categories based on the pathlines of the gas-particle mixture within the separator: uniflow cyclones and typical reverse flow cyclones [37, 38]. The uniflow cyclones enable the gas-particle mixture to flow in a single direction through a solitary channel, reducing the pressure difference between the inlet and outlet. The mixture enters the cyclone and swirls around the barrel body, completing the flow vortex in less than one revolution before exiting through the central bottom pipe of the cyclone. In the meantime, particles accumulate at the bottom of the cyclone's wall, with a small percentage (1–3%) of the inlet gas flow escaping along with the particles to improve collection efficiency and minimize pressure drop. This design offers a short-contact residence time, making it suitable for applications requiring high-solid loading and high-temperature, such as ultra-rapid fluidized reactors [35, 39, 40]. This design offers a short-contact residence time, making it suitable for applications requiring high-solid loading and high-temperature, such as ultra-rapid fluidized reactors [39, 41]. However, despite their small size, simple structure, and resulting low-pressure loss, uniflow cyclones have a relatively low particle collection efficiency [42].

In contrast, reverse-flow cyclone separators are the preferred choice of cyclone due to their high efficiency in separating gas-particle mixtures. However, they are also associated with significant body friction loss owing to their pronounced swirl effect [43]. In Chap. 7, this book describes the numerical simulations and the performance of the centrifugal reverse flow cyclone separator, as shown in Fig. 2.4, which is widely recognized as the most popular model in the field of separation and purification technology.

Reverse flow cyclones are ingenious devices that harness the power of centrifugal and gravitational forces to effectively separate solid particles from an air stream [44, 45]. A typical cyclone comprises of a barrel and conic sections that are topped off with an outlet tube. When the granular flow enters the cyclone through the inlet, it is set into a swirling motion that moves toward the conical part. As the granular flow reaches the conical part, centrifugal forces that effectively separate the solid particles from the air stream are generated. The separated particles fall into the bottom tip of the cyclone, while the air stream continues through the inner air vortex and exits through the cyclone outlet. Several factors must be considered, including the domain geometry, particle properties and interactions, inlet air velocity, and operating conditions to achieve high-performance separation in cyclone separators [44–47]. While cyclones have been primarily designed for fine particle control [48–50], no studies have been conducted on their use in controlling large particles,

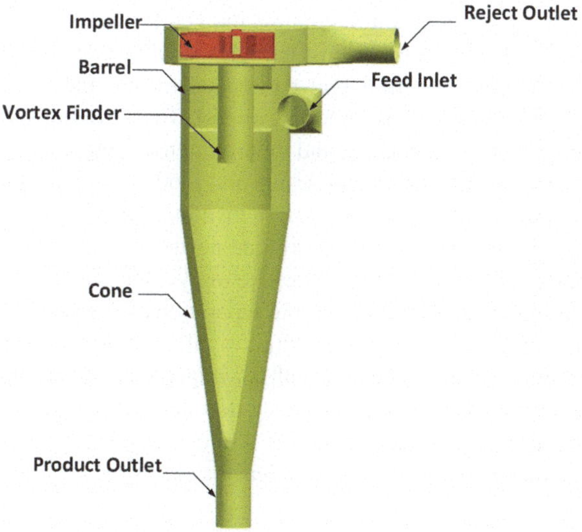

Fig. 2.4 Typical 3D view of the components of a reverse flow cyclone separator

particularly in the bio-processing industry. This presents an exciting opportunity for researchers to explore the potential of cyclone separators in this field. Optimizing cyclone design and operation parameters could pave the way for the more efficient and cost-effective separation of large particles in the bio-processing industry.

In the world of cyclone design, an innovative approach has emerged to facilitate easy comparisons between different models. Instead of using absolute sizes, the dimensions of various cyclone parts are expressed as a fraction concerning the cyclone diameter, making it possible to evaluate designs side-by-side. In the agricultural processing industry, the 1D3D [48] and 2D2D [49] cyclone models have gained popularity for their ability to separate fine particulate matter effectively. Additionally, the 1D2D model [50], known as the low-pressure cyclone, was created to replace covered condenser drums and is ideal for axial-flow fan exhaust applications [51]. The numbers preceding the "D's" in each model refer to the cylindrical and conical section lengths, respectively, while the "D's" indicate the cyclone diameter. For example, the 1D3D and 1D2D cyclones have cylindrical lengths equivalent to the cyclone diameter, and conical lengths of three and two times the cyclone diameter, respectively. The 2D2D cyclone, on the other hand, has cylindrical and conical lengths that are twice the cyclone diameter. With this innovative method of expressing cyclone design, the possibilities for enhancing particulate matter control are endless.

The 1D3D cyclone, a popular type of cyclone separator, typically features a traditional inlet design that is tall and narrow and another inlet that is modified to a 2D2D design. Recent studies have shown that incorporating the 2D2D inlet design and increasing the outlet diameter (tail cone) from D/4 to D/3 can yield significant benefits, including reduced emissions and back pressure and improved cyclone separation efficiency [52, 53]. Numerous experimental and numerical investigations have been conducted to better understand the flow dynamics within standard cyclone separators, yielding valuable insights into the design and operation of these devices [52, 54, 55]. While the 1D2D cyclone is well-suited for processing high-lint trash concentration in cotton ginning, the 1D3D and 2D2D cyclones are the most influential collectors for fine dust particles smaller than 100 μm in diameter and can also be effective for the larger particulate matter under certain conditions, albeit with greater energy consumption [56–58]. These findings demonstrate the potential for improved performance and environmental outcomes in cyclone separator applications through thoughtful design and modification of inlet and outlet geometries.

Admittedly, various research institutes have increasingly recognized the unparalleled efficiency of 1D3D cyclone design in numerous abatement systems, particularly for gin usage [46]. The size of cyclone dimensions concerning particle size is crucial in determining optimal separation efficiency. In earlier studies that employed the 2D2D design cyclone with a small diameter, all particles greater than 20 μm in diameter were collected with an impressive 100% separation efficiency [59, 60]. Moreover, it has been observed that the performance of cyclones can be enhanced by increasing inlet velocities while non-linearly decreasing with an increase in diameter [61]. Different efficiencies are expected since cyclones typically separate various particulate matter sizes. Coarse particulate matter is typically separated consistently, while fine materials tend to evade separation. Regardless of the type of cyclone utilized, it is critical to ensure proper sizing according to the airflow rate, appropriate configuration for the intended application, and proper construction and maintenance to ensure optimal efficiency [62]. By embracing 1D3D cyclone design, research institutes can achieve unprecedented results in abatement systems, especially for gin used. Proper consideration of cyclone dimensions and particle size can optimize separation efficiency. Improved performance can also be achieved by manipulating inlet velocities and diameter. Despite the range of particulate matter sizes, achieving varying degrees of separation efficiency is still possible. Cyclones can operate optimally by ensuring correct sizing, configuration, and maintenance, providing a reliable means of separating the particulate matter.

On the other hand, however, not all cyclones are created equal, and the design of a cyclone depends on its intended use [63]. Table 2.1 provides an overview of these standard designs and their uses. The dimension ratios of cyclones vary depending on their purpose. Cyclones designed for high efficiency typically have smaller inlet ducts than those designed for high flow. This is because high-efficiency designs require a more concentrated vortex to separate particles from gas effectively. To achieve this, the vortex finder tube dimensions are typically smaller, and the

Table 2.1 Standard dimensions of common aero-type reverse flow cyclone separator

	Geometry and component							
Cyclone type	Barrel diameter (*D*)	Barrel length (L_b)	Cone length (L_c)	Vortex length (L_v)	Vortex diameter (D_v)	Inlet height (*H*)	Inlet width (*W*)	Duct outlet (D_o)
1D2D (general purpose) [50]	D	D	2D	9/8D	1.6D	1/2D	1/4D	1/2D
2D2D (general purpose) [49]	D	2D	2D	5/8D	1/2D	1/2D	1/4D	1/4D
1D3D (general purpose) [48]	D	D	3D	9/8D	1/2D	D	1/8D	1/3D
Stairmand (high-efficiency) [64, 66]	D	3/2D	5/2D	1/2D	1/2D	1/2D	1/5D	3/8D
Stairmand (high-flow) [64, 66]	D	3/2D	5/2D	7/8D	3/4D	3/4D	3/8D	3/8D
Swift (high-efficiency) [65, 66]	D	1.4D	5/2D	1/2D	2/5D	0.44D	0.21D	2/5D
Swift (high-flow) [65, 66]	D	1.7D	2D	0.85D	3/4D	4/5D	0.35D	2/5D
Lapple (general purpose) [66, 76]	D	2D	2D	5/8D	1/2D	1/2D	1/4D	1/4D

height of the vortex finder is usually greater than the inlet height. This prevents short-circuiting of gas, where it bypasses the vortex formation and exits directly from the inlet to the outlet. While general purpose cyclones are designed to balance efficiency and flow, achieving higher efficiency often comes with a trade-off of increased flow [64, 65]. In fact, the high-efficiency cyclone of standard design can only process about half the amount of gas that a high-throughput design of the same diameter and overall height can process while operating at the same pressure drop [66]. This highlights the fact that there is no one-size-fits-all cyclone design. Researchers have therefore focused on redesigning various cyclone geometries to enhance their performance. These modifications include vortex finder shape and size changes [67–69], cyclone diameter [55, 70], inlet angle [71–73], and cyclone height [74, 75].

2.6 Summary

Using aerodynamic systems for separating, handling, and transporting seeds and grains has proven highly efficient and precise. These systems, whether vertical, horizontal, inclined, or centrifugal, rely on the distinct aerodynamic properties of particles to achieve adequate separation while maintaining the integrity of the final product. Vertical aerodynamic systems have shown promise in separating various grains like soybeans, corn, oats, wheat, and rye with remarkable efficiency. By carefully adjusting factors like straw length and injection methods, these systems can achieve high separation rates without compromising the quality of the grains. Additionally, innovative approaches like oscillating sieve devices have been instrumental in overcoming challenges such as particle clustering within the air stream. Horizontal aerodynamic systems offer a straightforward method for separating mixtures into different fractions. These systems have shown effectiveness in separating grains from weed and other non-grain materials, albeit with some challenges when dealing with small-sized grains. The inclined aerodynamic systems have proven valuable for sorting straw, grain, and chaff mixtures in combined harvesters. These systems can achieve maximum separation efficiency while minimizing grain loss by optimizing parameters such as the inclination angle and airflow velocity. The centrifugal aerodynamic systems, mainly cyclone separators, have been long-standing solutions for separating suspended particles from gas streams. With their simple yet effective design, these devices offer high separation efficiency, are widely used in various industries, and have significantly improved separation efficiency and environmental outcomes. Refining these systems continually through research and innovation can enhance efficiency, minimize waste, and improve the overall quality of agricultural products. The key lies in understanding and optimizing the parameters that influence separation efficiency, paving the way for sustainable and effective agricultural practices.

References

1. Sosulski F (1987) Yield and functional properties of air classified protein and starch fractions from eight legume flours. J Am Chem Soc 6:363–370
2. Uhl JB, Lamp BJ (1966) Pneumatic separation of grain and straw mixtures. Trans ASAE 61–135:244–246
3. Hollatz B, Quick GR (2003) Combine tailings returns, part 1: the effects of combine performance and settings on tailings. In: International conference on crop harvesting and processing. ASABE, St. Joseph, Louisville, p 22
4. Tylek P, Walczyk J (2003) Critical air velocity as a separation feature in nuts of European beech (Fagus sylvatica L.). Electron J Polish Agric Univ 6(2):4
5. IAEE (2008) Develop technical and technological bases improve separation of grain and seed set of physical and mechanical properties. Glevaha, IAEE
6. Gregg ES, Colton J, Matin MA, Krupnik TJ (2020) Efficient and participatory design of scale-appropriate agricultural machinery workshops in developing countries: a case study in Bangladesh. Dev Eng 5:100046. https://doi.org/10.1016/j.deveng.2019.100046
7. Stepanenko S, Kotov B, Kuzmych A, et al (2022) To the theory of grain motion in an uneven air flow in a vertical pneumatic separation channel with an annular cross section. Processes 10:1929. https://doi.org/10.3390/pr10101929
8. Stepanenko S (2017) Research pneumatic gravity separation grain materials. In: Scientific proceedings V international scientific-technical conference "agricultural machinery", Glevakha, Ukraine, pp 143–145
9. Reichert RD (1982) Air classification of peas (Pisum sativum) varying widely in protein content. J Food Sci 47:1263–1267
10. Emami S, Tabil LG, Tyler RT, Crerar WJ (2007) Starch-protein separation from chickpea flour using a hydrocyclone. J Food Eng 82:460–465. https://doi.org/10.1016/j.jfoodeng.2007.03.002
11. Panasiewicz M, Sobczak P, Mazur J et al (2012) The technique and analysis of the process of separation and cleaning grain materials. J Food Eng 109:603–608. https://doi.org/10.1016/j.jfoodeng.2011.10.010
12. Hellevang KJ (1985) Pneumatic grain conveyors. North Dakota State University, Fargo
13. Piven VV (2018) Determination of the extent of fraction in air separation of grain material. J Phys Conf Ser 1059. https://doi.org/10.1088/1742-6596/1059/1/012001
14. Shahbazi F (2013) Aerodynamic properties of wild mustard (Sinapis arvensis L.) seed for separation from canola. J Sci Food Agric 93:1466–1470. https://doi.org/10.1002/jsfa.5916
15. Gorial BY, O'Callaghan JR (1991) Separation of grain from straw in a vertical air stream. J Agric Eng Res 48:111–122. https://doi.org/10.1016/0021-8634(91)80008-3
16. Tylek P, Walczyk J (2004) Mechanical separation of seeds. Sci Bull 14:140–145
17. Farran IG, Macmillan RH (1979) Grain-chaff separation in a vertical air stream. J Agric Eng Res 24:115–129. https://doi.org/10.1016/0021-8634(79)90046-5
18. Gorial BY, O'Callaghan JR (1990) Aerodynamic properties of grain/straw materials. J Agric Eng Res 46:275–290. https://doi.org/10.1016/S0021-8634(05)80132-5
19. Tylek P, Walczyk J (2004) Effectiveness of the pneumatic separation of Norway spruce Picea abies (L.) karst seeds. Dendrobiology 51:101–104
20. Arinze EA, Schoenau GJ, Sokhansanj S, Adapa P (2003) Aerodynamic separation and fractional drying of alfalfa leaves and stems – a review and new concept. Dry Technol 21:1669–1698. https://doi.org/10.1081/DRT-120025503
21. Gan-Mor S, Wiseblum A, Regev R (1986) Separation of leaves from stems with a perforated rotating drum under suction. J Agric Eng Res 34:275–284. https://doi.org/10.1016/0021-8634(86)90036-3
22. Gorial BY, O'callaghan JR (1991) Separation of particles in a horizontal air stream. J Agric Eng Res 49:273–284. https://doi.org/10.1016/0021-8634(91)80044-F
23. Bilanski WK, Graham WD, Mowat DN, Mkomwa SS (1989) Separation of alfalfa silage into stem and leaf fractions in a horizontal airstream. Trans ASAE 32:1684–1690

24. Menzies D, Bilanski WK (1968) Aerodynamic properties of alfalfa particles. Trans ASAE 11:829–831
25. Bilanski WK, Jones D, Mowat DN (1986) Mechanical and aerodynamic separation of whole-plant corn silage into grain and stover. Trans ASAE 29:1187–1192
26. Simonyan KJ, Yiljep YD (2008) Investigating grain separation and cleaning efficiency distribution of a conventional stationary rasp-bar sorghum thresher. Agric Eng Int CIGR J X:1–13
27. Rothaug S, Wacker P, Yin W, Kutzbach HD (2003) Capacity increase of cleaning units by circular oscillation. In: International conference on crop harvesting and processing. ASABE, St. Joseph, Louisville, p 7
28. MacAulay JT, Lee JHA (1969) Grain separation on oscillating combine sieves as affected by material entrance conditions. Trans ASAE 12:0648–0651. https://doi.org/10.13031/2013.38919
29. Misener GC, Lee JHA (1973) Aerodynamic separation of grain from straw and chaff in a dispersed stream. Can Agric Eng 15:62–65
30. Pariset E, Parent C, Fouillet Y et al (2018) Separation of biological particles in a modular platform of cascaded deterministic lateral displacement modules. Sci Rep 8:1–10. https://doi.org/10.1038/s41598-018-34958-8
31. Rezvanivand Fanayi A, Nikbakht AM (2015) A CFD study of the effects of feed diameter on the pressure drop in acyclone separator. Int J Food Eng 11:71–77. https://doi.org/10.1515/ijfe-2014-0125
32. Yuu S, Jotaki T, Tomita Y, Yoshida K (1978) The reduction of pressure drop due to dust loading in a conventional cyclone. Chem Eng Sci 33:1573–1580. https://doi.org/10.1016/0009-2509(78)85132-X
33. Bing-tao Z (2006) Effects of flow parameters and inlet geometry on cyclone efficiency. Chin J Process Eng 6:178–180
34. Elsayed K, Lacor C (2011) The effect of cyclone inlet dimensions on the flow pattern and performance. Appl Math Model 35:1952–1968. https://doi.org/10.1016/j.compfluid.2012.09.027
35. Knowlton TM (2003) Cyclone separators. In: Yang W-C (ed) Handbook of fluid-particle systems, 2nd edn. Taylor & Francis Group LLC, New York, pp 603–620
36. Gimbun J, Chuah TG, Fakhru'l-Razi A, Choong TSY (2005) The influence of temperature and inlet velocity on cyclone pressure drop: a CFD study. Chem Eng Process 44:7–12. https://doi.org/10.1016/j.cep.2004.03.005
37. Oh J, Choi S, Kim J (2015) Numerical simulation of an internal flow field in a uniflow cyclone separator. Powder Technol 274:135–145. https://doi.org/10.1016/j.powtec.2015.01.015
38. Dehdarinejad E, Bayareh M (2021) An overview of numerical simulations on gas-solid cyclone separators with tangential inlet. ChemBioEng Rev 8:375–391. https://doi.org/10.1002/cben.202000034
39. Gauthier TA, Briens CL, Bergougnou MA, Galtier P (1990) Uniflow cyclone efficiency study. Powder Technol 62:217–225. https://doi.org/10.1016/0032-5910(90)80108-B
40. Sumner RJ, Briens CL, Bergougnou MA (1987) Study of a novel uniflow cyclone design. Can J Chem Eng 65:470–475. https://doi.org/10.1002/cjce.5450650316
41. Pillei M, Kofler T, Wierschem A, Kraxner M (2020) Intensification of uniflow cyclone performance at low loading. Powder Technol 360:522–533. https://doi.org/10.1016/j.powtec.2019.09.011
42. Oh J, Choi S, Kim J et al (2014) Particle separation with the concept of uniflow cyclone. Powder Technol 254:500–507. https://doi.org/10.1016/j.powtec.2014.01.057
43. Muschelknautz U, Pattis P, Kraxner M, Reinalter M (2011) Design criteria of uniflow cyclones for the separation of solid particles from gas. In: 10th international Conference on circulating fluidized beds and fluidization technology – CFB-10, p 8
44. Taiwo MI, Namadi MA, Mokwa B (2016) Design and analysis of cyclone dust separator. Am J Eng Res 5:130–134
45. Paraschiv SL, Paraschiv S (2016) Analysis of cyclone collection efficiency. Termotehnica 1:108–111

46. Wang L, Parnell CB, Shaw BW (2002) Performance characteristics of cyclones in cotton—gin dust removal. CIGR J Sci Res Dev IV:1–18
47. Ibhadode O, Ogedengbe E, Rosen M (2017) Performance characterization of gas-solid cyclone for separation of particle from syngas produced from food waste gasifier plant. Eur J Sustain Dev Res 13:1–14. https://doi.org/10.20897/ejosdr.201713
48. Parnell CBJ, Davis DD (1979) Predicted effects of the use of new cyclone designs on agricultural processing particle emissions. In: Southwest region meeting of the ASAE. ASAE, Hot Springs
49. Shepherd CB, Lapple CE (1939) Flow pattern and pressure drop. Ind Eng Chem 31:972–984. https://doi.org/10.1021/ie50356a012
50. Simpson SL, Parnell C (1995) New cyclone design for cotton gin emissions containing lint fiber. In: Dugger CP, Richter DA (eds) Proceedings beltwide cotton conference. National Cotton Council, Memphis, San Antonio, pp 680–687
51. Simpson SL (1996) Performance characteristics of a low pressure cyclone for axial-flow fan exhausts. Texas A&M University, College Station
52. Baker RV, Hughs SE (1999) Influence of air inlet and outlet design and trash exit size on 1d3d cyclone performance. Trans ASAE 42:17–21. https://doi.org/10.13031/2013.13193
53. Baker RV, Hughs SE, Gillum MN, Green JK (1996) Influence of trash exit design on cyclone performance. In: Beltwide cotton conference. National Cotton Council, Memphis, pp 1605–1609
54. Safikhani H, Akhavan-behabadi MA, Shams M, Rahimyan MH (2010) Numerical simulation of flow field in three types of standard cyclone separators. Adv Powder Technol 21:435–442. https://doi.org/10.1016/j.apt.2010.01.002
55. Faulkner W, Buser M, Whitelock D et al (2006) Effects of cyclone diameter on performance of 1D3D cyclones: collection efficiency. In: ASABE annual international meeting. American Society of Agricultural and Biological Engineers (ASABE), Portland, pp 1–13
56. Wang L, Parnell CB, Shaw BW (1999) Performance characteristics for the 1D2D, 2D2D, 1D3D and Barrel cyclones. In: ASAE annual international meeting, Paper No. 99-4195. ASAE, Toronto
57. Wang L (2000) A new engineering approach to cyclone design for cotton gins. Texas A&M University, College Station
58. Lingjuan W, Parnell CB, Shaw BW (2002) A study of the cyclone fractional efficiency curves. Agric Eng Int CIGR J Sci Res Dev IV:1–14
59. Wesley RA, Mayfield WD, MeCaskill OL (1972) An evaluation of the cyclone collector for cotton gins. United States Department of Agriculture, Economic Research Service, Washington, DC
60. Anthony WS, Mayfield WD (1994) Abatement of air pollution and disposal of gin waste. In: Anthony WS, Mayfield WD (eds) Cotton ginner's handbook, 503rd edn. USDA ARS, Washington, D.C., pp 175–182
61. Faulkner WB, Shaw BW (2006) Efficiency and pressure drop of cyclones across a range of inlet velocities. Appl Eng Agric 22:155–161
62. Whitelock DP, Armijo CB, Buser MD, Hughs SE (2009) Using cyclones effectively at cotton gins. Appl Eng Agric 25:563–576
63. Prasanna N, Subramanian K, Ajay S et al (2021) CFD study on the performance of reducing pressure drop holes in cyclone separator. Mater Today Proc 43:1960–1968. https://doi.org/10.1016/j.matpr.2020.11.425
64. Stairmand C (1951) The design and performance of cyclone separators. Trans Inst Chem 29:356–383
65. Swift P (1969) Dust control in industry. Steam Heat Eng 38:453–456
66. Leith D, Mehta DDI (1973) Cyclone performance and design. Atmos Environ 7:527–549. https://doi.org/10.1016/0004-6981(73)90006-1

67. Qiang L, Qinggong W, Weiwei X et al (2020) Experimental and computational analysis of a cyclone separator with a novel vortex finder. Powder Technol 360:398–410. https://doi.org/10.1016/j.powtec.2019.10.073
68. Brar LS, Sharma RP, Dwivedi R (2015) Effect of vortex finder diameter on flow field and collection efficiency of cyclone separators. Part Sci Technol 33:34–40. https://doi.org/10.1080/02726351.2014.933144
69. Brar LS, Elsayed K (2018) Analysis and optimization of cyclone separators with eccentric vortex finders using large eddy simulation and artificial neural network. Sep Purif Technol 207:269–283. https://doi.org/10.1016/j.seppur.2018.06.013
70. Brar LS, Sharma RP (2015) Effect of varying diameter on the performance of industrial scale gas cyclone dust separators. Mater Today Proc 2:3230–3237. https://doi.org/10.1016/j.matpr.2015.07.127
71. Wang S, Li H, Wang R et al (2019) Effect of the inlet angle on the performance of a cyclone separator using CFD-DEM. Adv Powder Technol 30:227–239. https://doi.org/10.1016/j.apt.2018.10.027
72. Zhao B, Shen H, Kang Y (2004) Development of a symmetrical spiral inlet to improve cyclone separator performance. Powder Technol 145:47–50. https://doi.org/10.1016/j.powtec.2004.06.001
73. Mageshwaran G, Durai Raj RB, Joseph GB et al (2018) Computational investigation of flow parameters and efficiency of single and double inlet cyclone separators. Int J Ambient Energy 39:707–712. https://doi.org/10.1080/01430750.2017.1303644
74. Hoffmann AC, De Groot M, Peng W et al (2001) Advantages and risks in increasing cyclone separator length. AICHE J 47:2452–2460. https://doi.org/10.1002/aic.690471109
75. Surmen A, Avci A, Karamangil MI (2011) Prediction of the maximum-efficiency cyclone length for a cyclone with a tangential entry. Powder Technol 207:1–8. https://doi.org/10.1016/j.powtec.2010.10.002
76. Lapple C (1951) Process use many collector types. Chem Eng 58:144–151

Chapter 3
Modeling of Aerodynamic Systems

Abstract This chapter describes the modeling philosophy of aerodynamic separation systems and their wide range of applications. These systems represent integral innovations across industries, from agriculture to pharmaceuticals, providing precise and efficient methods for sorting and handling particles based on their aerodynamic properties. These systems rely on a complex interplay of fluid dynamics, particle behavior, and engineering design. This chapter examines the fundamental principles of aerodynamic separation laying the groundwork for understanding how aerodynamic separation devices are designed and operated. The theoretical framework, computational modeling, and real-world applications reveal the complex movements of biogenic particles such as seeds and grains, as they navigate these advanced separation systems. In addition, the chapter highlights the crucial frontier of the interplay between the biogenic particles and the fluid dynamics that dictates their behavior. The chapter also delves into the methodologies of modeling these particles, shedding light on their varied sources, behaviors, and impacts within the separation processes. The comprehensive overview provides insight into the modeling approaches, challenges, and opportunities that define the realm of aerodynamic separation systems and their diverse agricultural applications.

Keywords Aerodynamic systems · Computational simulation · Granular material · Multi-sphere approach

3.1 Relevant Aerodynamic Concepts

Aerodynamics is the fascinating study of how gases, particularly air, interact with moving solid objects. By examining the forces of lift and drag generated by air flow over and around solid bodies, aerodynamics has become a crucial factor in numerous industries. For instance, forced air technology in industrial processing can be used to efficiently separate components with unique aerodynamic properties from the main product. Conversely, if no such differences exist, alternative methods are employed. In the agricultural sector, up to 70% of impurities in post-harvest grain

L. Zhou et al., *Discrete Element Method for Multiphase Flows with Biogenic Particles*, https://doi.org/10.1007/978-3-031-67729-8_3

material are separated from the main crop by utilizing their distinct aerodynamic characteristics [1]. When it comes to machines that use air, the efficiency of their operation depends heavily on the weight and aerodynamic drag of the particles they are handling. This, in turn, influences their terminal velocity of fall [2], highlighting the necessity of designing aerodynamic equipment with a keen awareness of the properties of the materials being processed. In the pursuit of optimizing this process, researchers have measured the drag coefficients of a wide range of agricultural particles by subjecting them to airstreams and analyzing their suspension velocities [3, 4]. The Reynolds number has been utilized to better understand the relationship between drag coefficients and particle shape [5]. By grouping particles based on their spherical or cylindrical shape, one can improve the ability to predict the terminal velocity of grains and other materials accurately.

When understanding particles' aerodynamic properties, there are two key factors to consider, namely, the critical or terminal velocity and the drag coefficient [6]. The first parameter, which considers the main representation of particles' aerodynamic properties, can be mainly used as their distinguishing feature in such aerodynamic applications and optimizations [4, 7, 8]. The terminal velocity is a critical representation of a particle's aerodynamic properties. It refers to the velocity of the vertical air stream that allows the particle to remain suspended in the air by balancing its weight with the force of the air stream [9–11]. This parameter is influenced by secondary factors such as density, shape, and cross-sectional area [6], which can be calculated using Eq. (3.1) [12, 13]. Interestingly, the particle's mass and area significantly impact the terminal velocity. A particle with a greater mass has a higher terminal velocity, while a larger particle area results in a lower velocity.

$$V_T = \sqrt{\frac{m_i g}{k_a A_i \rho_a}} \text{ or } \cong \sqrt{\frac{4d_i\left(\rho_i - \rho_a\right)g}{3C_d \rho_a}} \tag{3.1}$$

where V_T is the terminal velocity, m_i is the mass of the particle, g is the gravity acceleration, k_a is the dimensionless coefficient of aerodynamic resistance, A_i is the projected or lifting area of the particle, d_i is the particle diameter, ρ_i and ρ_a are the particle and air density, C_d is the drag coefficient.

The drag coefficient is a dimensionless particle parameter used to quantify how easy or difficult it is to entrust a particle to fluid flow [14]. It is related to the particle Reynolds number, which is given by Eq. (3.2).

$$R_{e_i} = \frac{\rho_a d_i V_T}{\mu_a} \tag{3.2}$$

where R_{e_i} is the particle Reynolds number and μ_a is the dynamic viscosity of the air.

The terminal velocity of particles is a crucial factor in determining the particle Reynolds number. However, obtaining an accurate calculation of terminal velocity is a complex and iterative process. To better understand the behavior of particles in fluid flows, three distinct regimes have been identified based on their Reynolds number: Stokes' regime when $R_{e_i} < 0.2$, the intermediate regime when $0.2 R_{e_i} 1000$, and the Newton regime when $R_{e_i} > 1000$ that are described in detail in the following section. When dealing with spherical and cylindrical particles in the range of Reynolds numbers of $R_{e_i} = 1000$ to 200,000, a common assumption is that the drag coefficient C_d is 0.44 and 1.0, respectively [6, 15]. However, experimental data has shown that the exact value can vary depending on the grain material and other conditions [9, 16–19]. To accurately determine the terminal velocity of the particles, researchers have utilized a suspension apparatus test, which typically involves a rotating impeller and a fine wire grid screen, as shown in Fig. 3.1. By delivering a uniform air velocity field through the screen and impeller, it can measure the airflow velocity when the sample reaches the flotation point, where the rotational movement of the particles is at its lowest. Using an air velocity meter provides accurate data for determining the terminal velocity of the particles. When it comes to optimizing the aerodynamics of agricultural cereal grain crops' aerodynamics, expressing each particle's projected area can be quite a challenge, owing to their irregular shapes. However, one approach that has proven to be effective is assuming that the particles are spherical and using the diameter of a sphere with the same volume as the particle's designed size as a reference point. This method is beneficial when

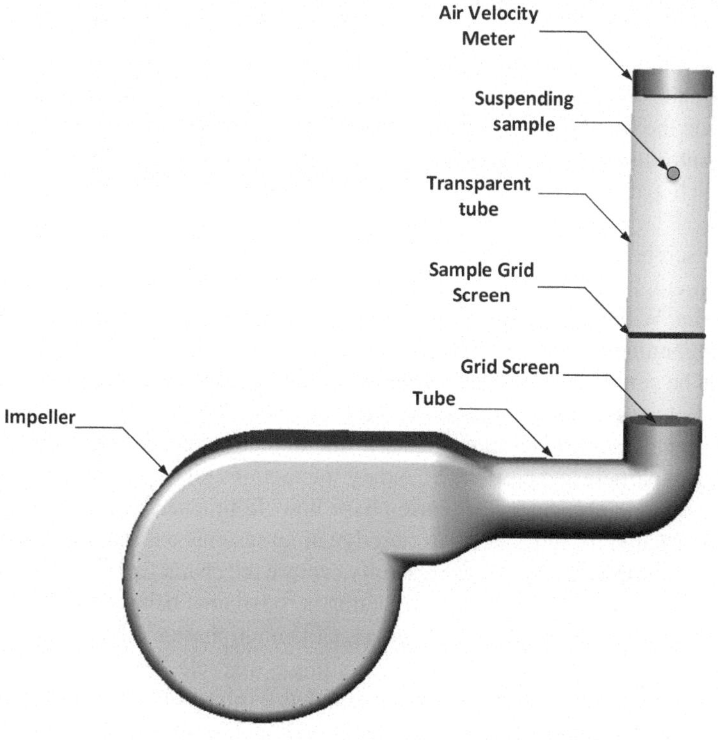

Fig. 3.1 Schematic diagram of the apparatus for measuring terminal velocity

evaluating particles with complex shapes, where determining the volume is difficult. Alternatively, one can also use the geometric mean diameter, approximating the diameter of the equivalent sphere, provided the shape factor is close to unity [11].

Theoretically, the net force acting on a particle moving in a fluid is defined by the difference between the drag and the gravitational forces as given in Eq. (3.3) [5].

$$F_{\text{net}} = \frac{1}{2} C_d A_i \rho_a V_a^2 - m_i g \tag{3.3}$$

where F_{net} is the net force acting on the particle and V_a is the velocity of the air.

3.2 Modeling Philosophy

As mentioned earlier, in recent years, enhancing the performance and design of harvesting and post-harvesting devices such as grain dryers, spouted beds, aerodynamic separators, and combine harvesters has become a pressing need. To achieve this goal, using experimental studies, mathematical models, and computer simulations has become an indispensable tool. With the advent of cutting-edge technology and increased computing power, computational simulations have proven to be highly reliable and adaptable for accurately mimicking operational conditions. In particular, numerical models with relevant modifications have been employed in previous and recent studies to describe the intricate particulate matter and gas flow field inside the aerodynamic systems [20–26]. Since the bulk bio-particulate matter mass is introduced to the structure domain, it is considered a discrete element, where each particle is an element having an interaction with the neighboring elements.

The discrete element method (DEM) is a powerful numerical simulation technique that employs Newton's laws of motion to predict and model the flow of discrete media. With the widespread use of agricultural particulate matter in industrial processing, numerous articles have been published using DEM [27–35]. Many particles can be simulated, thanks to computers, in a straightforward manner with significant memories and high-performance processors. By tracking each particle's rotational and translational movements in the domain, considering contact forces and particle–particle and particle–boundary interactions [36, 37], DEM can dynamically simulate the trajectory of dense phase flow. In practical applications, where particulate matter is introduced to aerodynamic systems via air or gas flow, it becomes a multiphase flow. To successfully capture the characteristics of air or fluid flow and obtain important data for predicting aerodynamic behavior [38–41], CFD models are commonly employed. However, CFD multiphase models are inadequate for estimating the particulate matter phases in gas–particle flows due to the many restrictions that arise when dense flow is applied. Fortunately, coupling CFD with DEM can overcome these restrictions [42], which is described in detail in Chaps. 4

and 5. Although this coupled process increases computational time, it enables a resolved particle-particle interaction combined with gas-to-solid and solid-to-gas modeled interactions.

3.3 Geometry Components and Parameters

Aerodynamic systems play a crucial role in the agricultural industry, especially in the transportation and separation of biogenic particles. To accurately model and predict the flow dynamics of these particles, a combination of mathematical and numerical approaches have been utilized, including theoretical frameworks, differential equations, incremental techniques, and computer programming. One notable example of such modeling is the work of Adewumi [43], who employed MATLAB to simulate the flow of cowpea grains in a horizontal airstream. Adewumi developed a 2D model based on aerodynamic principles, considering drag and gravitational forces. The remarkable results showcased the ability to achieve separation, cleaning, sorting, grading, and density classification processes within a single machine. In recent years, the advancement of computer technology has enabled the utilization of the CFD–DEM coupling approach for modeling and predicting air–particle flow problems in agriculture [44, 45]. This approach has become increasingly popular due to its ability to provide comprehensive solutions and insights into the behavior of particles in aerodynamic systems, which are summarized in the following sections. In conclusion, the integration of mathematical and numerical models, along with the power of computer technology, has revolutionized the field of aerodynamics in agriculture. The CFD–DEM coupling approach has provided valuable insights and solutions for optimizing bio-particle flow dynamics in aerodynamic systems, leading to more efficient and effective agricultural processes. With further advancements in modeling techniques, even more innovative solutions are expected for the future. Thus, whether designing a new agricultural system or optimizing an existing one, incorporating these advanced modeling approaches can significantly enhance outcomes and drive success in the industry.

3.4 Air-Screening System

Effective separation and cleaning of grain materials from impurities, such as spikelets, leaves, straw remnants, husks, and small weed seeds, is crucial in modern agricultural grain production. Advanced harvesting machines, like combine harvesters, utilize sophisticated air-screening or air-sieving systems to achieve this process [46]. The intricate dynamics of this system can be numerically described as a complex two-phase flow involving the interplay between the airflow generated by a centrifugal fan and the dispersion of solid particles within the system by the airflow [47]. As previously mentioned, air velocity plays a pivotal role in this separation process. The airflow carries away the lighter particles due to the aerodynamic drag force, while the heavier ones are primarily separated by the force of gravity and the inertia of sieves. The delicate balance between these forces is essential in achieving optimal separation and cleaning results, ensuring that only pure, high-quality grains are harvested.

In recent research by Li et al. [23], a 3D-coupled CFD–DEM approach was employed to investigate the impact of inlet airflow velocity on the separation of rice grains and short straws in an air-and-screen cleaning unit. The study considered various factors, including longitudinal velocity, vertical height, and cleaning loss. It utilized EDEM software to model the rice grains as spheroid particles and the short straws as cylinder shapes. These particles were then allowed to fall onto an inclined vibrating screen, with an Eulerian–Eulerian model used to simulate the airflow using Ansys Fluent® as part of the coupling. The Hertz-Mindlin contact model was applied to replicate particle–particle and particle–screen (wall) collisions accurately. The simulation results reported that the screen length could be shortened if material straws and foreign materials were too small. The higher the inlet airflow velocity, the faster the backward velocity of the materials on the screen surface, and the greater the cleaning loss. The simulation results revealed that the screen length could be optimized by managing the presence of material straws and foreign objects. Specifically, it was observed that higher inlet airflow velocities resulted in faster backward velocities of materials on the screen surface, leading to increased cleaning losses. This critical finding highlights the need to control inlet airflow velocities to optimize cleaning efficiency carefully.

Building on these findings, Li et al. [48] further extended the research to investigate the effects of throughput and operating parameters on cleaning performance in the same device. Ansys Fluent® was utilized as an auxiliary investigation method to analyze the movement behavior of the internal airflow. The standard $k - \varepsilon$ turbulence model with Reynolds-averaged Navier-Stokes (RANS) equations of fluid dynamics was proposed for this simulation study to accurately capture the real physical processes at different airflow velocities above and below the vibrating screen. The results, as depicted in Fig. 3.2, demonstrated that the fan speed significantly influenced cleaning performance by affecting the overall airflow velocities in the cleaning device, while the throughput had a pronounced effect on the grain impurity ratio.

Furthermore, Lenaerts et al. [49] used DEM simulations to model a combined harvester's separation chamber. They represented grains as spherical particles and straw particles as segmented hollow cylinders that can be bent in all directions using DEMeter++ software. They found that particle properties influence separation speed, with straw coverage and grain diameter having the main effects. Ma et al. [50] developed a variable-amplitude screen box to analyze the reciprocating screen in a combined harvester. Using a multi-sphere approach, they used ellipsoidal particles to represent rice grains and cylindrical particles to represent rice straw. The simulation showed that the variable-amplitude screen caused particles at the front of the screen to be quickly thrown up and moved back but did not promote segregation and separation at the screen front based on the Hertz-Mindlin contact model in

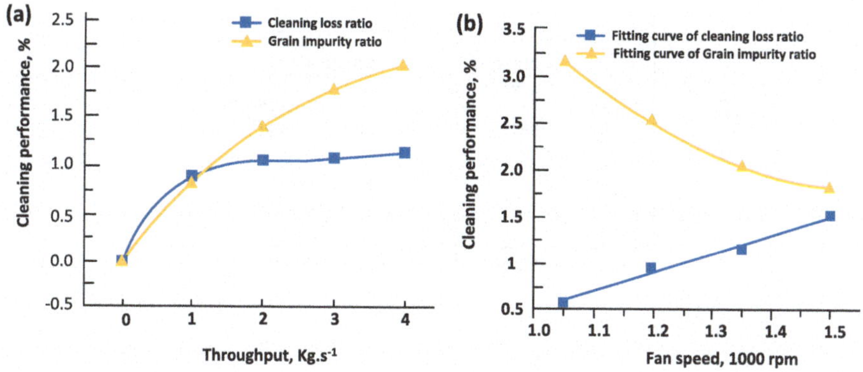

Fig. 3.2 (**a**) Polynomial diagram of throughput and (**b**) fan speed with cleaning performance [48]

EDEM software. So, the main findings were related to the effects of particle properties and screen design on the separation process in combined harvesters. Overall, these studies provide insights into the mechanics of grain straw separation in combined harvesters and the factors that influence this process.

On the other hand, Enchen et al. [51] studied the grain behavior in an inertia separation chamber of the combined harvester based on CFD–DEM simulation. The grains were modeled as ellipsoidal, while the straw was modeled as cylindrical shapes through the multi-sphere approach. The $k-\varepsilon$ turbulence model was used in Ansys Fluent® software to reach the convergence state of the airflow, while the Eulerian–Lagrangian method in EDEM code was applied to model the particles. The simulations showed that the grains were effectively separated from the short straw through the airflow. Besides, the performance of separating and cleaning grains improved with the increase in airflow velocity, but turbulence and pressure drop were seriously increased at the end of the separation chamber.

Yuan et al. [52] investigated the screening process of rice-threshed mixture components (grains, shriveled grains, and short stalks) through a cylindrical sieve device using CFD–DEM coupling. The influence of inlet airflow velocity, diameter, and the cylindrical sieve's rotating speed on the separation process was studied. First, the particles were scanned by a 3D laser scanning system, then modeled by the multi-balls fast filling method (API) in EDEM software and allowed to move smoothly through the cylinder sieve. The particle phase is considered in a continuous quasi-fluid phase and solved through the averaged Navier-Stokes equation, where the airflow was simulated through the existing Eulerian–Eulerian model in the Ansys Fluent® code. Figure 3.3 visualizes the simulation of the separation process and the particle trajectories with the time through the device, where the yellow, green, and blue colors represent grains, shriveled grains, and short stalks, respectively. The numerical simulation showed increased screening efficiency when the cylindrical sieve diameter increased. At the same time, the rotating speed significantly affects the average axial velocity of the short stalks, screening loss rate, and screening efficiency.

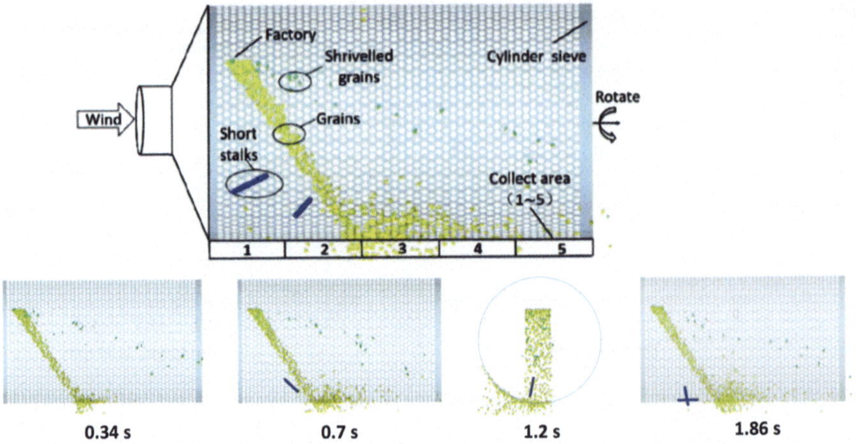

Fig. 3.3 The particle trajectories of the rice-threshed mixture during the simulation process [52]

As shown in the investigations, the technique and methodology of the numerical simulations used for modeling the air-screening systems can provide a good reference for predicting the flow behavior and solving the separation problems of agricultural materials. Because CFD–DEM coupling models had a good prediction compared to the experimental results in the above investigations, they are strongly recommended to be used for improving and optimizing the design of air-screening devices in the combined harvesters.

3.5 Seed-Metering Device

Precision agriculture has revolutionized agricultural production and management, making them more accurate and effective [53]. Among the various aspects of precision agriculture, precision seeding has emerged as a key focus of researchers [54]. It involves the precise placement of individual seeds in the soil, ensuring optimal plant spacing [55]. Central to this process is the seed-metering device, a critical component of the planter machine that plays a pivotal role in achieving superior seeding performance. In recent times, aerodynamic seed-metering devices have gained widespread popularity due to their exceptional accuracy, ability to preserve seed integrity, and suitability for high-speed operations [56, 57]. Researchers have successfully modeled the air-seed flow in aerodynamic seed-metering devices, leveraging innovative numerical simulation approaches and enhancing their effectiveness and performance.

For instance, Dandan et al. [58] utilized the CFD–DEM coupling method to investigate the influence of different air nozzle installation positions on the working efficiency of the inside-filling air-blowing precision seed-metering device. They concluded that this simulation approach could successfully detect and predict the

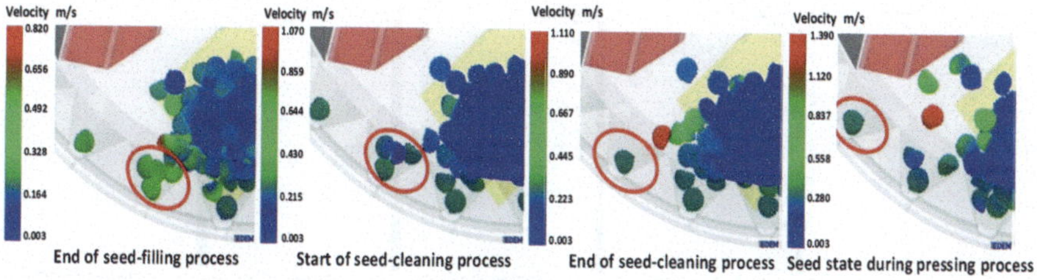

Fig. 3.4 Seed motion behavior of the seed-cleaning process [58]

movement behavior of maize seeds in a single seed meter. Han et al. [26] extended the research to optimize the air-maize flow in the same device using the same numerical methodology. The team studied the impact of lateral hole positions, width, and average arc length on the gas field and seed movement. The maize particles were modeled as a hard sphere using EDEM software with the bonded particle model technique, which allowed the particles to be treated as a soft sphere model. In this study, the standard $k - \varepsilon$ two-equation turbulence model with the RANS equations in Ansys Fluent was used to solve the gas phase, and the Eulerian two-fluid coupling approach was employed. The simulations have shown that the motion of seeds during the seed-cleaning process can be accurately predicted, as demonstrated in Fig. 3.4. Additionally, the simulation results have indicated that the size of the lateral hole significantly impacts the movement and airflow of seeds within the hole. The optimal lateral hole has a width of 1.5 mm and an average arc length of 10 mm, and its position plays a crucial role in determining its effectiveness. This optimized hole performs exceptionally well when positioned lower, resulting in less pressure loss, superior drag force, and differential pressure. Experimental testing of the optimized seed-metering device has yielded a 93% success rate when working pressure exceeds 5.5 kPa and the device operates at a speed lower than 10 km/h.

Lei et al. [59] analyzed the impact of the structure of the throat, seed feed rate, and airflow inlet velocity on the behavior of air and seeds in a seed-feeding device called a pneumatic Venturi tube. This device is commonly used in pneumatic seeding planter machines. Using a CFD–DEM coupling approach, they modeled rapeseed and wheat as spheres and multi-spheres, respectively, using EDEM software. The particle modeling was done using the Hertz-Mindlin model, while the gas phase was solved using the standard $k - \varepsilon$ two-equation turbulence model with the RANS equations in Ansys Fluent. The Lagrangian coupling method was employed to implement the coupling interface. The simulation results indicated that the throat's area and airflow inlet velocity mainly impacted the airflow outlet velocity, while the length of the throat and seed feed rate slightly influenced seed movement. Increasing the throat area led to a decrease in seed velocity and pressure loss within a specific range. At the same time, an increase in inlet air velocity resulted in the development of resultant force and seed velocity. Figure 3.5 displays rapeseed and wheat seed simulation flow patterns at the intended air velocities. Thus, it can be inferred that the suitable airflow inlet velocity for rapeseed and wheat was 16–20 m/s and 20–24 m/s, respectively.

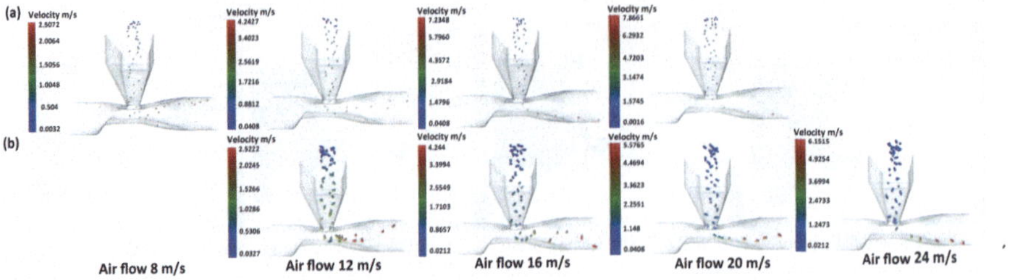

Fig. 3.5 Flow patterns of the particles at different air velocities (**a**) rapeseed and (**b**) wheat seed [59]

Despite the Venturi tube's widespread use in various industries, there remains a dearth of numerical studies exploring the behavior of particles in aerodynamic seeding systems. Lei et al. [60] have taken up this challenge and extended the frontiers of knowledge in this area. Their groundbreaking study sought to characterize the migration trajectory of rapeseed and wheat particles and predict their distribution behavior within a pneumatic distribution head in an air-assisted centralized seed-metering device (Planter). To achieve this feat, the researchers focused on key parameters, including the streamlined angle, baseline radius, outlet diameter of the streamlined angle, lid angle, and installation location within the distribution head. They also employed a hard-sphere model to represent the particles, with the Hertz-Mindlin (no-slip) model as the particle contact model. By adjusting the seed feed rating through EDEM software, the researchers could model the behavior of the rapeseed and wheat particles. They used a multi-sphere approach to generate ellipsoidal-shaped wheat particles while modeling rapeseed particles as spherical. The standard $k - \varepsilon$ turbulence model with RANS equations was employed to account for turbulence in the airflow, and the airflow velocity was set in the Ansys Fluent® software. The simulation results first predicted the variation coefficient of seed distribution in the head. Interestingly, it was observed that as the streamlined angle increased from 10° to 50° for an inlet diameter of 20 mm and an airflow velocity of 20 m/s, the variation decreased initially and then increased. When the radius of the baseline was less than 20 mm, the air pressure loss and resultant seed force also increased. Furthermore, the seed's resultant force and resultant velocity increased with the increase in airflow velocity (Fig. 3.6a, b). The study demonstrated that the structure of the streamlined tube significantly influenced the distribution behavior of seeds. When dispersed in the pressurized tube, the seeds initially moved upward at a lower velocity, progressing from the entrance of the distribution head to the outlet tube along the baseline. During this process, the seeds rotated with the lid, with many directly colliding with the surfaces of the shell and lid. Consequently, the seeds, initially moving in a vertical direction, shifted to a horizontal trajectory due to the combined effects of the airflow and the distribution head's structure. Notably, the distribution of seeds was most uniform when the streamlined angle was 60°, the baseline radius was 40 mm, and the lid angle was 120°, as shown in Fig. 3.6a. Furthermore, to enhance distribution uniformity and reduce power consumption, seeds could be efficiently separated into eight outlet tubes through the streamlined tube at airflow velocities of 20–24 m/s and 24–28 m/s for rapeseed and wheat seeds, respectively, as shown in Fig. 3.6b.

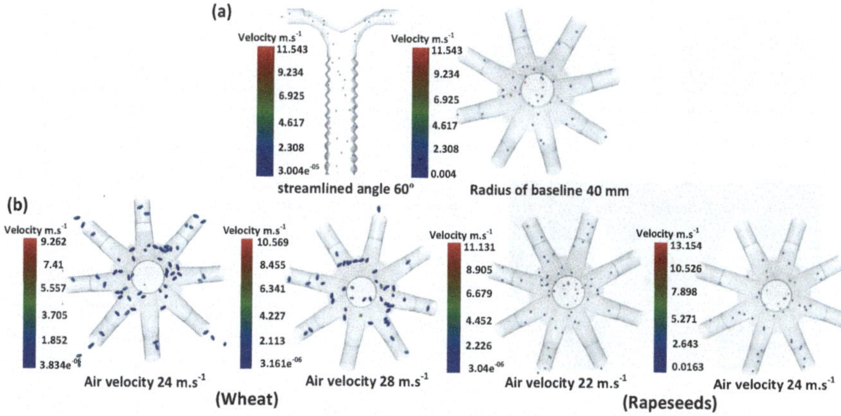

Fig. 3.6 Snapshots of wheat and rapeseeds simulation in the distribution head [60]

The findings unequivocally demonstrated that the CFD–DEM coupling methodology is a trustworthy and practical instrument for replicating the intricate interplay between seed motion and the airflow domain. The simulation outcomes hold immense potential for enhancing the evenness of seed dispersion, unraveling the underlying mechanisms of seed distribution, augmenting efficiency, and refining the blueprint of air-borne seed dispensers.

3.6 Drying Systems

Drying technology is pivotal in preserving food grains, seeds, and kernels, ensuring their longevity after harvesting [61, 62]. Continuous drying is a widespread technique in the agricultural industry that relies on mixed-flow or cross-flow dryers (MFDs). These dryers leverage forced dry air convection through a grain bed using air ducts, preventing microbial activity and spoilage [63]. To optimize MFD performance, Keppler et al. [64] explored the impact of constructional modifications on air duct shape and wall friction. The study employed a multi-sphere approach to model wheat particles as a clump of three spheres and used the Hertz-Mindlin contact model to simulate interaction forces through the EDEM code package. The researchers evaluated two modified air duct shapes, namely, wedge and circular ducts, as shown in Fig. 3.7a, b. The simulation results revealed a significant discrepancy in velocity distribution using circular air ducts, which is unsuitable for wheat drying applications. Additionally, particle placement within the module considerably impacted velocity, with near-wall velocities significantly lower than those in the middle, as depicted in Fig. 3.7c, d. The study recommends using smooth structural materials to reduce the particle–wall friction ratio, leading to a more uniform velocity distribution. By employing this approach, it becomes easier to achieve homogeneous drying, ensuring optimal preservation of particles.

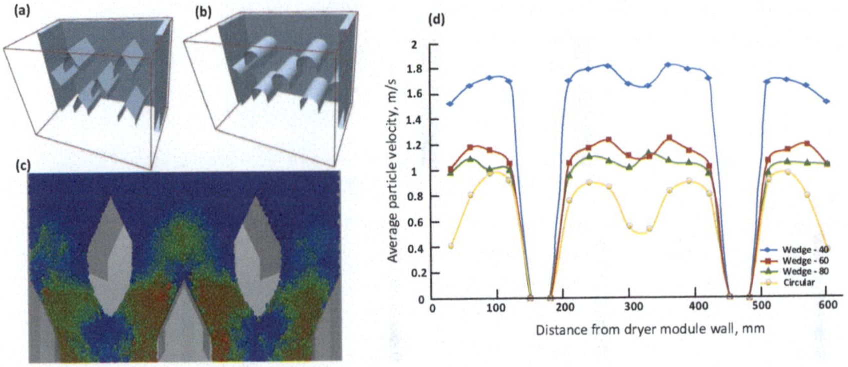

Fig. 3.7 Simulated velocity differences around the different air ducts (right): (**a**) wedge duct, (**b**) circular duct, (**c**) velocity distribution of the wheat (red denotes higher and blue the lower velocities), and (**d**) velocity differences from the dryer wall [64]

To avoid extra drying costs and improve product quality, Weigler et al. [65] set out to revolutionize traditional mixed-flow dryers to enhance product quality while reducing unnecessary drying expenses. Their approach involved utilizing CFD–DEM to scrutinize the dryer's design for potential flaws. The study explored various dryer geometries and air duct arrangements, including horizontal and diagonal configurations. The researchers chose wheat particles as the model material, and they employed particle flow code (DEM-PFC 2D) software to compare particle flow simulations using two geometries, namely, an ellipsoidal clump and a sphere. The airflow domain was discretized using ANSYS ICEM, and the airflow was modeled as a turbulent flow utilizing the two-equation shear stress transport (SST) model with Ansys CFX® software. The simulation results proved to be consistent with experimental findings, successfully capturing the particle–air flow distribution within the dryer. The researchers discovered that traditional mixed-flow dryers are far from perfect due to air duct arrangements that cause dead zones in airflow and an uneven distribution of airflow caused by wall frictions. They recommended implementing suitable air duct arrangements to regulate air velocities in the middle section and near the sidewalls of the dryer. Consequently, the researchers suggested a new dryer model incorporating inclined sidewalls and half-air duct removal to balance airflow distribution with particle flow velocities. So, Weigler et al. [66] went above and beyond in their pursuit of a more efficient and effective method for particle drying. Through their extensive research and experimentation, they have developed a revolutionary geometry for MFDs that promises to revolutionize the industry. By utilizing advanced technology such as DEM-PFC2D and transparent sidewalls, they could track wheat flow and observe the drying process in real-time. Their findings indicate that the new design yields more uniform drying and improves energy efficiency and overall product quality.

On the other hand, Weigler and Mellmann [61] made a noteworthy discovery in their research regarding the behavior of dry wheat kernels in the absence of airflow, where the results were compared with the experiments. They utilized a multi-sphere approach to create an ellipsoidal clump to model the wheat kernels numerically,

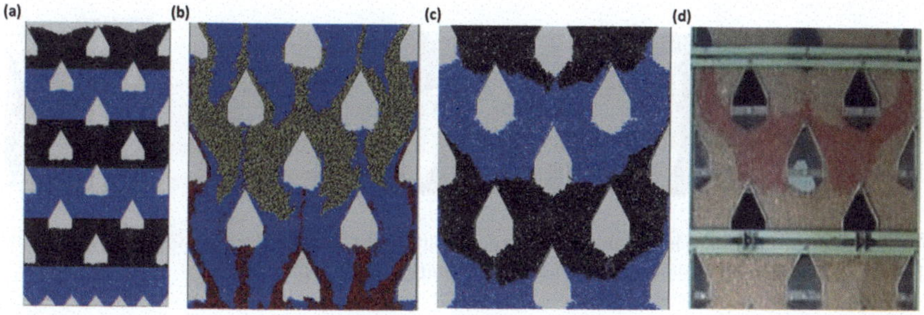

Fig. 3.8 Particle flow patterns in MFD: (**a**) particles after DEM gravitational generation, (**b**) DEM sphere particles, (**c**) DEM ellipsoidal clamp, and (**d**) experimental result [61]

which was then compared to the flow behavior of a spherical shape. To better observe the mass flow behavior, the particles were colored black and blue and viewed in a 2D setting, as depicted in Fig. 3.8. The results were fascinating, as the DEM model utilizing the ellipsoidal particle could accurately predict the actual grain flow behavior. This was evident in the experiment, where the colored grain layer displayed the same profile as the numerical simulation. On the other hand, when spherical particles were used, the bed angle typically formed under air ducts was absent.

The remarkable similarity between the simulation results and the experimental observations shed light on the predominant flow characteristics of particle movement within the dryer. Specifically, the central region of the dryer exhibits higher flow velocities with reduced frictional effects, while the particles traversing the areas proximal to the dryer walls experience greater friction and consequently move with lower vertical velocities. This differential flow behavior engenders non-uniform moisture distribution at the dryer outlet, underscoring the need for novel design strategies to mitigate variations in particle residence time. To this end, a three-dimensional simulation is recommended for more accurate grain mass flow patterns and velocity predictions, as the 2D-DEM model cannot accurately estimate absolute values in this context.

In an exciting breakthrough, Khatchatourian et al. [21] took the initiative to create a cutting-edge 3D model of the flow of soybeans. They utilized the advanced DEM-Yade software package to simulate and forecast the motion and behavior of particles during discharge in MFDs. The particles were cleverly modeled as a single spherical geometry in 3D, considering their physical properties. To validate their simulation data, they performed an experiment, which yielded exceptional agreement with their results, as exemplified in the visually striking Fig. 3.9a.

Using the colored particle technique to analyze continuous flow in the MFD, distinct variations in the velocities of soybeans could be discerned. Notably, seed velocities were lower near the wall regions than the center, as illustrated in Fig. 3.9b. This non-uniform velocity field translates into differential residence times for the particles, a finding that corroborates earlier studies. To optimize flow in dryers, the authors propose an integration of mass and heat transfer models with the air-particle flow, which promises to yield more accurate simulation results.

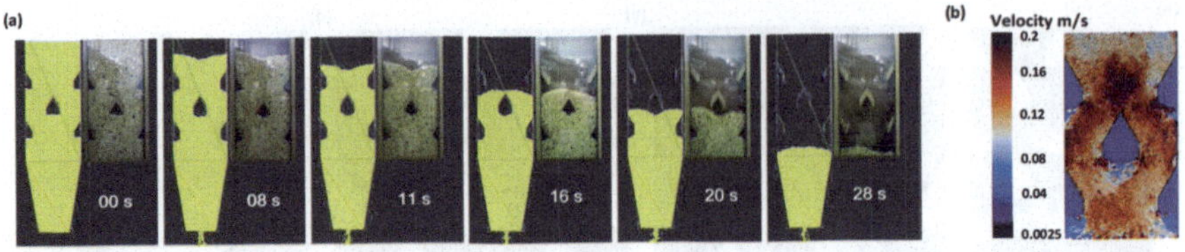

Fig. 3.9 (**a**) Simulation (left frame) and experiment (right frame) comparison of soybean discharge at a discharging hole of 20 mm and (**b**) velocity field distribution [21]

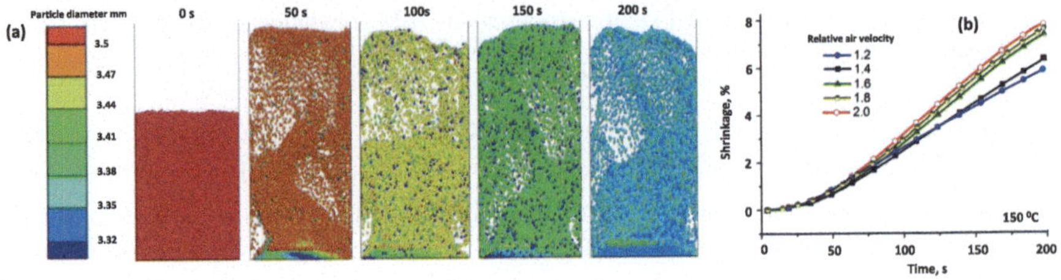

Fig. 3.10 (**a**) Snapshot of particle size change throughout the drying process and (**b**) the effect of inlet air velocity on particle size reduction [68]

As a result, Scaar et al. [67] extended the work on the developed MFD model recommended by Weigler et al. [65] using the heat and mass transfer models. Their study aims to predict the airflow distribution and investigate the effect of air duct arrangements. Based on the 3D CFD Ansys CFX® code, two different bed materials, namely, wheat and rapeseed, were used and considered as a porous medium. The turbulence of fluid flow is modeled by an eddy viscosity approach using $SST - k - \omega$ model with RANS equations in Ansys CFX. The numerical observations were validated by experimental data, which showed that the air duct arrangement has a strong influence on the airflow distribution. In addition, the horizontal air duct arrangement showed a more homogeneous airflow pattern as compared to the diagonal one. Furthermore, the numerical model calculated quite well the influence of the bed material with its different particle characteristics, which gives good support for using this model in future investigations.

On the other hand, Azmir et al. [68] have devised a bespoke CFD–DEM drying model for food grains that considers the grain's shrinkage properties. They aimed to examine how inlet air temperature and velocity affect the drying process in a fluidized bed. To achieve this, they used wheat particles with a spherical shape as their model grain and coupled a 2D CFD with a 3D DEM. They assumed that there was only one control volume in the thickness direction and that the source in the bed thickness direction was insignificant. To dry the grains, they used hot air with temperatures ranging from 100° to 180° and a given velocity of 1.2–2 m/s, introduced from the two corners of the bed. The drying model they developed could accurately predict the overall particle shrinkage rate and the diameter of each particle during the simulation time, as shown in Fig. 3.10. Their findings indicated that the particle shrinkage rate increased significantly with an increase in inlet air temperature but only slightly with an increase in inlet air velocity, within the range studied. Overall, this study sheds light on the intricate interplay between drying parameters and particle shrinkage, which has important implications for optimizing the drying process of food grains.

In their quest to delve deeper into the drying process, Azmir et al. [69] conducted a groundbreaking study examining grain properties' impact, including size, density, and initial moisture content, on the process. Employing the same cutting-edge CFD–DEM coupling methodology, they focused on millet particles, comparing them with five other species of spherical particles. By subjecting the grains to three different inlet air temperatures (50, 150, and 200 °C), they sought to gain insights into the various heat transfer modes and their effects. The numerical simulations unearthed that convection and conduction played a significant role in the drying process, while radiation had a negligible impact. Moreover, they discovered that as the grain size decreased, the drying rate experienced an exponential increase. However, this decrease in size also reduced the uniformity of moisture content and size distribution. Similarly, reducing the initial moisture content resulted in a lower drying rate, leading to greater homogeneity in the moisture content. Lastly, they observed that increasing the grain density slowed the drying rate, resulting in longer drying times and lower moisture content uniformity. In summary, Azmir and his team's findings provide valuable insights into the drying process, which can have far-reaching implications for the agricultural and food industries.

As a sequel, computational modeling, particularly the use of CFD–DEM coupling techniques, offers a plethora of benefits when it comes to grain drying. One of the most significant advantages is its ability to examine grain distribution and air velocity throughout the dryer sections, where changes can influence geometry. With the insights gained from previous findings, coupling this approach with mass and heat transfer models can be immensely advantageous when dealing with realistic industrial-scale drying processes. This modeling technique enables us to decipher the complex behavior of grains within the dryer, thereby eliminating the need for costly prototypes.

3.7 Pneumatic Separation Systems

In the realm of process engineering, the separation process plays a vital role in refining raw materials to obtain more homogeneous fractions of substances based on size or chemical characteristics. While fluidization simulations for mineral particles and medicines have been extensively studied to improve equipment development and process efficiency, the realm of agricultural particulate matter remains largely unexplored. However, a recent study by Almeida et al. [36], as shown in Fig. 3.11, has made strides in this field by developing a new computational model to separate sugarcane bagasse particles pneumatically using a riser device based on CFD Ansys Fluent® and DEM tools. The linear hysteresis model was applied to simulate the interaction between particles and boundaries, and the Lagrangian–Eulerian method was utilized as a one-way coupling approach. In addition, the $k - \varepsilon$ turbulence model with the RANS equations was chosen in Ansys Fluent to account for air

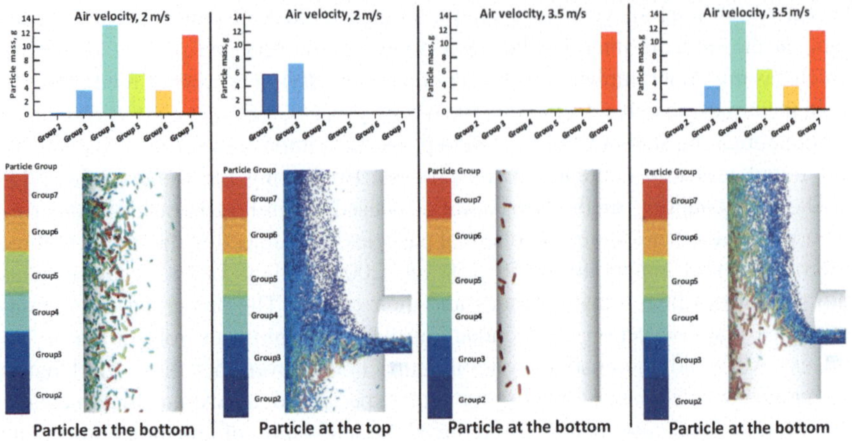

Fig. 3.11 Particle mass distributions at the bottom and top of the riser device with different air velocities [36]

turbulence flow. The particles were classified into seven groups based on their characteristics and aspect ratio to simulate the separation process. Group 1 represents the fine pith particles, Groups 2–4 represent the coarse fibers, and Groups 5–7 represent the coarse-grained (CG) particles. The fibers and rind particles were modeled as cylinders. In contrast, the pith particles were omitted from the simulation as the air would entrain them due to the chosen velocities. Including them would consume excessive computational power and time. The researchers chose air velocities of 2.0 m/s and 3.5 m/s based on the calculated terminal velocities of the particles. The resulting particle mass distributions at the top and bottom of the riser device were analyzed and visualized for each air velocity, revealing that the coarse particles could be easily tracked during separation, even with similar properties.

Advancements in one-way and two-way coupling approaches have yielded promising results. A recent study by the same previous researchers [14] aimed to establish a theoretical foundation and gain insight into separation phenomena to enhance the equipment used in particle separation processes. Using the same particle modeling methodology as a previous study, the authors collected data on the mass distribution of dry and wet bagasse at the top and bottom outlets of the riser under varying air velocities. The two coupling approaches employed provided highly accurate predictions and demonstrated the potential for further use in drying, pyrolysis, fluidization, and gasification processes, paving the way for the development of optimal device configurations.

Meanwhile, gas cyclones remain a prevalent method of pneumatic separation in many industrial applications, which removes dust particles from gases or recovers valuable products. However, achieving high performance in gas cyclones requires a thorough understanding of the underlying physics of the process, which has proven to be a challenging task. Early empirical models failed to account for critical factors significantly impacting cyclone performance, including unsteadiness, asymmetry, high turbulence, and reversal flow [70, 71]. Additionally, traditional mathematical models struggle to predict performance under high swirls and turbulent conditions accurately, hindering their effectiveness in industrial applications [72, 73]. Despite these challenges, continued research and innovation hold promise for further optimizing gas cyclones and unlocking their full potential.

Gas cyclones have traditionally been simulated using CFD numerical models. However, the standard $k - \varepsilon$ turbulence model used in these models can lead to excessive turbulent eddy viscosities and unrealistic tangential velocities when simulating flows with a swirl. Boysan et al. [74] were among the first to develop a CFD model for gas cyclones, and their findings suggested that turbulence models such as the Reynolds stress model (RSM) and large eddy simulation (LES), which can deal with anisotropic turbulence problems, should be used instead. Recent studies have applied these turbulence models to gas cyclones, and their numerical results have shown good agreement with experimental measurements [75, 76]. In the past, the Lagrangian particle tracking (LPT) method was mainly used for simulating particle flow, with the effect of solids on gas flow and particle–particle interaction force being largely ignored, except for a few continuum models [77]. While the LPT method can qualitatively study the effect of parameters such as geometry, gas flow velocity, and particle size, it falls short in describing the effect of solids on gas flow and parameters such as solid concentration or loading ratio. As a result, various investigators have made efforts to overcome this limitation. Overall, gas cyclone simulations require advanced numerical models that can accurately account for the complex interactions between gas and solid phases. With the emergence of turbulence models, such as RSM and LES, and continued efforts to improve particle flow simulations, the accuracy and reliability of numerical models for gas cyclones are continuously improving.

Previously, Xiang and Lee [78] conducted a study investigating the flow pattern in five cyclones at varying barrel heights. The objective was to determine the separation efficiency using the commercial CFD Ansys Fluent package. The study modeled the flow as steady and incompressible, with the turbulence simulated through the RSM models in the RANS equations. The findings from the study were quite intriguing. One notable discovery was that the tangential velocity drastically decreased as the cyclone height increased. This phenomenon, however, was not reflected in the separation efficiency. In addition, it was observed that when the gas exit tube received low flow amounts, the separation efficiency remained unaffected due to the higher tangential velocity. The study further recommended that the vortex finder design should be such that it does not protrude into the conical part. This way, the flow shortcuts are minimized, resulting in higher efficiency. In a groundbreaking study by Wang et al. [79], the numerical simulation of gas–particle flow in a typical Lapple cyclone was presented. To achieve higher accuracy, the researchers utilized the RSM model to simulate anisotropic turbulent flow and the Lagrangian approach to predict particle flow behavior. Impressively, the team simulated 150,000 particles with diameters ranging from 2×10^{-4} to 0.1 m. Their research revealed fascinating insights, such as the significant effect of particle position at the cyclone inlet and the lower separation efficiency at the bottom of the entrance. The study also demonstrated the impact of different particle sizes on behavior within the cyclone, as depicted in Fig. 3.12. Intriguingly, particles larger than the critical diameter

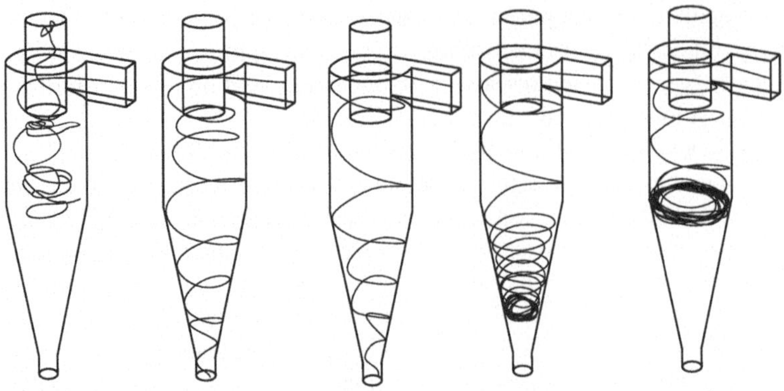

Fig. 3.12 The trajectories of particles in the cyclone with different diameters [79]

(1×10^{-5} m) were found to stagnate on the conical body of the cyclone, owing to the delicate balance between the supporting force from the wall, the gas drag force, and the gravity force.

Narasimha et al. [80] employed Ansys Fluent to conduct a multiphase simulation of turbulent-driven flow in a dense medium cyclone to predict particle distribution. Using the algebraic slip mixture multiphase model, they applied LES and RSM to model dispersed phases and turbulence closure, respectively. Their study found that ultra-fine particle sizes were distributed uniformly throughout the cyclone, whereas larger sizes were segregated close to the wall. Bernardo et al. [81] proposed a modified industrial-sized cyclone for gas–particle flow by adjusting the normal scroll inlet section angle to 30°, 45°, and 60°. They used the Ansys-CFX code to evaluate collection efficiency and pressure drop, employing the Lagrangian–Eulerian formulation to simulate numerically. RSM was used to model anisotropic flow in the cyclone, with cement particles of an average diameter of 26 μm. Results showed that total pressure drop decreased with increasing inlet section angle, and collection efficiency improved by 40% when using an inlet section angle of 45°. This proposed inlet angle represents a practical option for achieving a considerable power reduction in separation processes. Zhao et al. [82] simulated gas–particle flow using the Eulerian–Lagrangian approach through CFD Fluent to characterize the performance of two-type Stairmand cyclone double spiral inlets and a conventional single inlet. The study used the Lagrangian method to track limestone particle motion with a diameter of 7.14 μm and calculate gas–particle separation efficiency. The RSM was employed to model gas flow turbulence. Numerical results revealed that adding a double spiral inlet can improve the symmetry of gas flow behavior and enhance particle separation efficiency.

De Brito Dias et al. [83] conducted a study using Ansys CFX to simulate the multiphase flow of cyclones, using two different approaches, namely, Eulerian–Eulerian (E–E) and Eulerian–(Eulerian)$_n$ (E–E$_n$). The E–E approach considered only one particulate matter phase (talcum powder) with a mean diameter, while the

$E–E_n$ approach represented different phases of particulate matter sizes. The RSM was utilized to model the complex anisotropic flow turbulence, where different inlet velocities were introduced. The performance of the cyclone was assessed by its collection efficiency, which was found to be higher using the $E–E_n$ approach. It was also noted that the $E–E_n$ approach is faster, more scalable, and provides greater accuracy, making it an ideal alternative for obtaining cyclone collection efficiency. Meier et al. [84] continued the research on an $E–E_3$ code based on the CFD algorithm and Eulerian–Eulerian approach, allowing the modeling of up to three solid phases of talcum powder, each with a specific volumetric fraction, density, and particle size in addition to the gas phase in the cyclone separator. The solid phase was treated as a hypothetical gas, and the drag forces between phases were responsible for the gas–solid interaction. The gas turbulence was modeled using the $k - \varepsilon$ standard model and Prandtl's longitudinal mixing model (zero-equation model) with the RANS equations in Ansys Fluent. The numerical results showed that this quadruple model is more accurate and appropriate than the usual binary model for designing cyclones, improving their assembly efficiency, and reducing pressure. The quadruple model also provided the ability to visualize the dynamic flow inside the cyclone.

Gimbun [85] predicted the performance of a cyclone separator at extreme temperatures by utilizing Ansys Fluent. In their research, they employed the RSM for turbulence modeling and the Lagrangian frame using the discrete phase model (DPM) approach to study the behavior of particle diameters ranging from 0.6 to 10 μm. The operating temperature was modeled using the kinetic theory to predict the physical gas properties. The numerical simulations were conducted by coupling unsteady and steady states. The results revealed that the collection efficiency decreases as the temperature rises due to a decrease in tangential velocity. Chu et al. [38] conducted a numerical study of gas–particle flow in a Lapple gas cyclone using a CFD–DEM coupling approach. They solved the RANS equations for continuum gas flow, while Newton's equations of motion tracked the discrete particle phase. The research considered various gas-particle loading ratios from 0 to 2.5 with an interval of 0.5, considering both gas–particle and particle–particle interactions. The particle was assumed to be spherical with a diameter of 2 mm. As the particle loading ratio increased, the number of turns traveled by particles in a cyclone decreased, especially in the apex region, where the width of the strands increased, as shown in Fig. 3.13a. The tangential velocity of the gas phase decreased steadily, with the high axial velocity region moving upwards. The radial flow of the gas phase was significantly dampened, particularly in the apex region. The research also found more particles accumulated in the cyclone when the particle loading ratio was low. At the end of the simulation time, with a particle loading ratio of 2.5, the strand flow pattern was not observed in the conic section when using one-way coupling conditions. This phenomenon occurred because the reaction forces of the particles on the gas flow were not considered, in contrast to the two-way coupling, as shown in Fig. 3.13b.

In a study by Shukla et al. [86], the Eulerian–Lagrangian approach based on the LES and the RSM models was used to predict the collection efficiency of the cyclone under the effect of velocity fluctuations. The results revealed that the LES

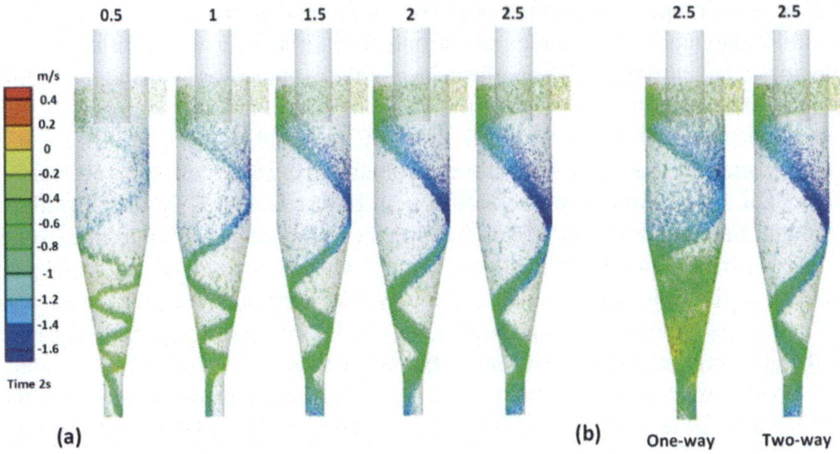

Fig. 3.13 Comparison of solid flow patterns in a typical Lapple gas cyclone; (**a**) different particle loading ratios and (**b**) one-way and two-way coupling at ratio 2.5 [38]

model outperformed the RSM model in predicting the fluctuating flow field and collection efficiency for each particle size, especially for small particles. The LES model's accuracy in predicting velocity fluctuations and the turbulent dispersion of particles was attributed to its ability to accurately resolve the vortex core around the cyclone symmetry axis as a three-dimensional time-dependent motion. Elsayed [87] optimized a new cyclone dimension for minimum pressure drop and compared its performance with the Stairmand cyclone separator using the Co-Kriging model based on CFD and the Muschelknautz model. The numerical data indicated that the new cyclone design closely resembled the Stairmand high-efficiency design in the geometrical parameter ratio, except for the cone height. The new cyclone design demonstrated superior performance in low-pressure drop, resulting in a 22% decrease in pressure drop compared to the Stairmand design at the same volume flow rate. Additionally, the study found that the inlet dimensions, vortex finder diameter, and cone height significantly improved the cyclone's performance, pressure drop, and flow field pattern. Kumar and Roy [88] investigated the effect of barrel diameter (D) on the performance parameters of the Stairmand high-efficiency cyclone. The study used the Ansys Fluent code and the RSM to control the chaotic behavior of the gas, considering dust particles with diameters ranging from 0.6 to 5.8 μm and tracked by Lagrangian methodology. The study found that the pressure drop and overall collection efficiency increase with increasing cyclone barrel diameter, while the cut-off diameter decreases.

Song et al. [89] employed the Eulerian–Lagrangian multiphase flow model in their study on the forces impacting limestone and the separation efficacy of a cyclone. The CFD code served as their tool of choice, while the LPT model was utilized to define the movement of each particle. The RSM was implemented to simulate the turbulent gas flow. The researchers assumed the injection of 11,200

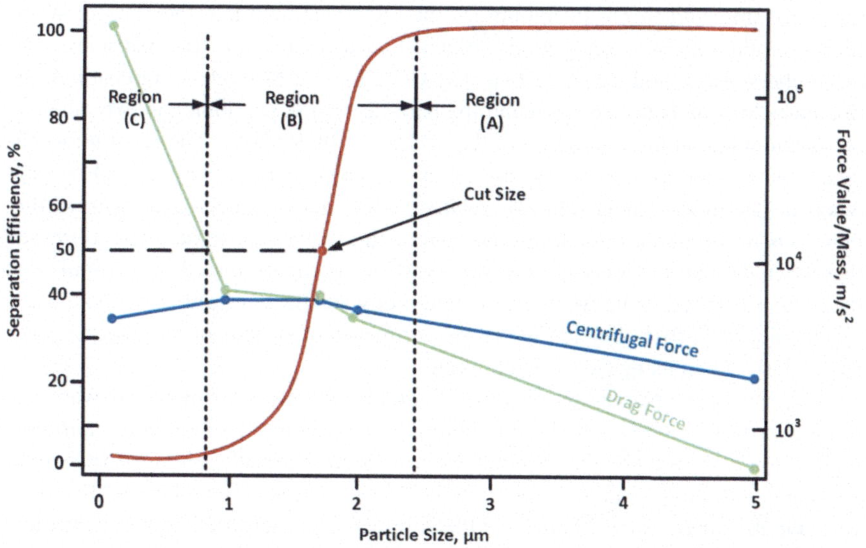

Fig. 3.14 Illustration of separation efficiency and average forces in the radial direction as a function of particle size [89]

particles for each particle size, varying from 0.2 to 5 μm in diameter. The numerical analysis results are encapsulated in Fig. 3.14, illustrating that large particles in Region A experience a significant centrifugal force compared to the drag force, leading to a 100% separation efficiency. However, as the particle size diminishes in Region B, the distinction between the centrifugal and drag forces becomes less pronounced until it reaches the cut size point, where the particles become highly susceptible to turbulence. As for particles in Region C, the drag force suddenly increases and surpasses the centrifugal force, rendering the particle motion exclusively reliant on the gas behavior.

In a fascinating study by Chu et al. [90], a CG CFD–DEM model was developed to simulate the complex swirling multiphase flow in a dense medium cyclone. The researchers compared the performance of their CG CFD–DEM model with the standard CFD–DEM model. While the former was generally accurate for dilute flows, minor discrepancies were observed for dense flows. Moreover, the CG CFD–DEM model outperformed the CFD–LPT model in modeling particle–particle and particle–gas interactions. These results highlight the potential of the CG CFD–DEM model as a valuable tool for quickly evaluating the flow and performance of large-scale dense medium cyclones. Interestingly, the computational time required for the CG CFD–DEM model was significantly lower than the standard CFD–DEM model, which can be attributed to the reduced particle count and increased time step for DEM simulation. Building on this success, Chu et al. continued their investigation in 2022, exploring the gas–particle physical flow features in a gas cyclone. Using the same modeling methodology, they studied the behavior of particles with fine and CG particle sizes ranging from 1–200 μm to 0.5–2 mm, respectively [91]. The results of this study were intriguing, as they demonstrated that the van der Waals forces could enhance separation efficiency, especially for fine particles. The researchers also found that their two-way CG CFD–DEM model could capture the fine particle strands flow pattern and the particle ceiling phenomenon, where particles accumulate at the top region of the cyclone cylinder body. The separation efficiency was found to increase slightly with an increase in the particle loading ratio when the particle loading rate was relatively low but then decreased with further increases in the particle loading ratio. Interestingly, the researchers also observed that when the particle loading ratio was very low, particles tended to accumulate at the intersection region between the cylinder body and the cone of the cyclone, while increasing the particle loading led to the formation of strands of flowing particles, particularly when using two-way coupling.

In recent years, Napolitano et al. [92] conducted a study using a four-way coupling approach of CFD and DEM to simulate the gas–particle flow in a Stairmand high-efficiency cyclone. The researchers compared the results obtained from their approach with those obtained from pure DEM simulations. The gas velocities were varied in the range of 10–20 m/s, and the volumetric particle loading was controlled between 0.1% and 0.5% with a diameter of 500 μm. The results showed that the quality of the solids' distribution at the walls of the cyclone is affected when using a low coarse-graining factor in pure DEM simulations. However, when the coarse-graining factor is increased, the gas and solid flow strands disappear in most cases. It was also found that DEM's results are more effective in determining slight inefficiencies in gas–particle flows than CFD's results. Kozołub et al. [93] conducted a numerical simulation of high particle loads inside a cyclone using the hybrid Eulerian–Lagrangian framework through the CFD–DDPM four-way coupling approach. The researchers used the RSM to account for turbulence effects, while the hybrid Eulerian–Lagrangian approach was used to model gas–particle behavior. The simulations were carried out for different mass loadings of spherical glass particles with diameters ranging from 40 to 80 μm and inlet velocities ranging from 6.94 to 9.49 m/s. The study revealed that particle flow significantly affects axial and tangential velocities reduction, resulting in weaker vortex rotation than in the gas phase. Additionally, the number of particle flow turnarounds that reach the cyclone ash hopper decreases for higher values of particle mass flow rates. The researchers suggested that more complex models considering the exact particle–particle and particle–wall interactions should be investigated further using different particle restitution coefficients.

Zhang et al. [94] explored the impact of different vortex finder diameters and depths of insertion on the pressure drop and separation efficiency of fine particles. The researchers used the RSM and DPM models in CFD Fluent software to simulate the gas flow turbulence and two-phase flow. The results showed that increasing the vortex finder diameter could reduce the total pressure drop and improve separation efficiency. Interestingly, changes in the length of the vortex finder had a more significant impact on the swirling flow field's velocity and structure than changes in the diameter. Wargadalam et al. [95] utilized the multiphase particle-in-cell (MP-PIC) method to model the gas–solid interaction in a small-scale cyclone. The

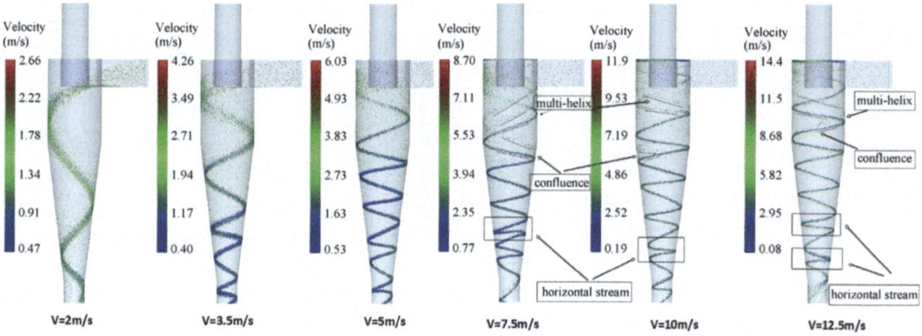

Fig. 3.15 Ultra-light particles pattern at different inlet gas velocities [96]

team employed the Eulerian–Lagrangian approach and CFD framework to simulate flue gas flow with catalyst particles ranging from 90 to 130 μm in diameter. Their study agreed well with analytical solutions, demonstrating the MP-PIC method's efficacy in simulating gas–solid flow. Zhou et al. [96] performed a numerical simulation to study ultra-light particle separation in a Stairmand cyclone at various inlet velocities. The researchers used the CFD–DEM coupling technique to describe the cyclone separation characteristics. The RSM model was applied to simulate turbulent gas flow, while Newton's second law through the DEM approach was used to track micron-sized graphite particles. The study revealed that the ultra-light particles did not impact the flow field and could be wholly separated even at low inlet velocities due to the small gas flow force acting on the particles. As the inlet velocity increased, the helix angle of the particle stream decreased, and the spiral number of the particles increased, leading to a denser and narrower particle string and a thinner radial particle layer, as shown in Fig. 3.15.

In a study by Wang et al. [97], the influence of the inlet angle on the cyclone's performance was simulated using CFD–DEM. Using Newton's second law of motion, this approach combined the RSM to characterize the turbulent gas flow and the DEM to track the particles. As depicted in Fig. 3.16, the results revealed that the particle size significantly affects the separation efficiency, which increases with increasing particle size. However, the inlet angle plays a crucial role in the performance, as the separation efficiency increases and then declines as the inlet angle changes from negative to positive. Interestingly, the maximum separation efficiency is achieved at an inlet angle of 15°. Moreover, the study found that the negative inlet angle causes a more significant pressure drop than the positive inlet angle. In terms of tangential velocity, the study showed that the inlet angle has only a slight effect, and only a more negative inlet angle would impact the overall tangential flow.

Nemati et al. [98] utilized CFD to model the behavior of saffron flower biogenic particles in a 2D2D cyclone separator, employing an Eulerian–Lagrangian approach based on Ansys Fluent software. By tracking discrete biogenic particles using the Lagrangian method and modeling the turbulence flow with the RSM, they could predict the separation efficiency of stigmas from petals with remarkable accuracy. In fact, their numerical results showed a separation efficiency of 89% at an inlet air

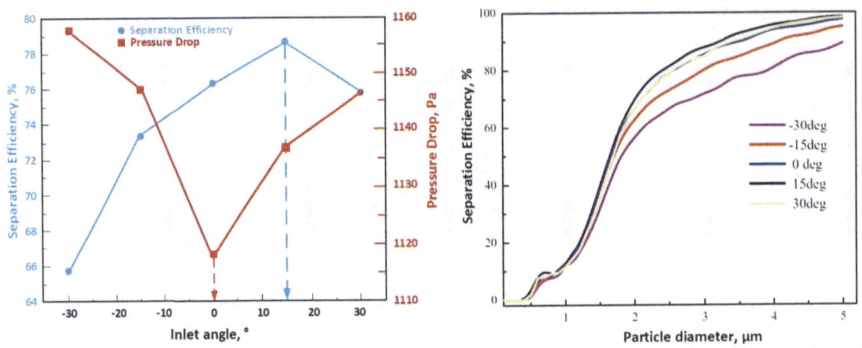

Fig. 3.16 Cyclone performance at different inlet angles [97]

velocity of 3.5 m/s, which closely matched experimental data. Therefore, the CFD simulation offers a reliable means of modeling biogenic particle matter and can effectively demonstrate gas-biogenic particle behavior. Chlebnikovas et al. [99] explored the potential of CFD in simulating aerodynamic processes in an innovative design for a multi-channel cyclone separator. By utilizing the modified turbulence models, including $k - \varepsilon$ and $k - \omega$ models with RANS equations, to simulate the anisotropic gas flow, they successfully modeled the gas cleaning phenomenon in the improved fourth-channel cyclone separator. This led to optimization of the research process and opened up new possibilities for using CFD in other gas-cleaning applications.

Mirzaei et al. [100] utilized a hybrid Eulerian–Lagrangian multiphase model, incorporating an agglomeration model and a sub-grid drag model using Ansys Fluent. This allowed the researchers to accurately predict the separation efficiency of a pilot-scale cyclone under high particle loading. The modeling of fine particles less than 15 μm was done using the DDPM approach, which captured the particle–particle interactions. The $k - \omega$ *SST* turbulence model with RANS equations was used to model the turbulent flow, which was sensitized to the rotation and curvature effects of the flow stream. The researchers found that the agglomeration model was crucial in predicting pressure drop and separation efficiency, and the sub-grid drag modification improved the prediction of separation efficiency. They also reported that the combination of these models was necessary to achieve accurate predictions of the separation efficiency, as the pressure drop was highly over-predicted without it. In another study by Morin et al. [101], the effect of solid loading and inlet aspect ratio on cyclone efficiency and pressure drop was investigated through CFD simulations. Glass beads with an average diameter of 42.2 μm were used as particles. The results showed that changing the inlet aspect ratio while keeping the inlet area constant did not affect the pressure drop and performance. However, at a higher inlet aspect ratio, the particle separation efficiency increased. The study recommended using pressure boundary conditions at the dipleg outlet as the best way to simulate cyclones in gas-sized bed circulation.

The review of the existing literature indicates that there has been a dearth of studies on the intricate interplay of gas and particle flow within a gas cyclone using a CFD–DEM approach incorporating the true shape of the particles in modeling. While prior literature has explored this topic to some extent, as evidenced by the works [14, 33, 76, 93, 102], the additional research reported in this book takes a step further by presenting a novel CFD–DEM model in Chap. 7 that accounts for the true shape modeling of the particles mixed with a gas in a gas cyclone separator. This new research aims to delve deeper into this complex gas–particle flow system by investigating the effect of particle loading ratio. By doing so, a comprehensive understanding of this system was generated, ultimately optimizing its performance. This book seeks to shed new light on this topic and contribute to the existing knowledge in this area.

3.8 Modeling of Biogenic Particles

Granular materials are widely used across various industries, from chemistry to civil engineering and agriculture. Despite their widespread use, the behavior of granular materials is still not fully understood. A pile of particles, for instance, can exhibit the characteristics of a porous solid yet flow like a liquid when placed on an inclined plane. Under significant vibrations, granular solids may even display gas-like properties. This phenomenon was first described in an ancient proverb dating back to 98–55 B.C., which likens the behavior of poppy seeds to that of water: "One can scoop up poppy seeds with a ladle as easily as if they were water and, when dipping the ladle, the seeds flow in a continuous stream" [103]. To model these materials accurately, their physical properties, such as size, density, and moisture content, must be carefully considered and determined individually for each DEM modeling application. The shape of the granular material alone cannot fully capture its complex behavior. Therefore, a detailed understanding of its physical properties is crucial for effective modeling. With this knowledge, scientists and engineers can harness the unique characteristics of granular materials to develop innovative and effective solutions across various industries [104].

Accurately simulating real-world processes relies heavily on realistic particle modeling in simulations [105–108]. For agricultural granular materials, which encompasses grains, seeds, tubers, and soft fruits, a diverse array of shapes is present, ranging from spherical (like rapeseed) to ellipsoidal (like rice) and diamond with round or sharp-edged (like corn or buckwheat) to highly complex, non-convex shapes of fruits and tubers [35]. These materials vary in size from less than a millimeter to several centimeters, making their modeling a significant challenge. While measuring the length, width, and thickness of the particles is crucial for modeling, describing the complex shapes necessitates the determination of the roundness of edges and the ratio of dimensions [104, 109]. Non-spherical shapes are common in actual particles, but they pose difficulties in representing their shapes in simulation software, making their use rare in simulation studies [105]. In earlier DEM simulations, particles were modeled as discs or spheres since normal contact forces always act toward the center, and particles' rotation is only influenced by tangential contact forces [110]. In an experimental study that investigated 14 particle geometries, including spheres, tetrahedra, cubes, octahedra, dodecahedra, icosahedra truncated, tetrahedra, triangular bipyramids, tetrahedral frames, and jacks with arm lengths, particle shape had a significant impact on granular material's mechanical response and interaction forces, particularly on stiffness and yielding parameters [111]. The particles' size, shape, and physical and mechanical properties pose significant challenges in operating DEM simulation modeling, as do the computing power, time consumption, and the number of discrete particles in the system [104, 111, 112].

Agricultural grains often come in complex shapes or non-spherical particles that pose a challenge for modeling in the DEM [105, 113–115]. To overcome this challenge, researchers have developed the multi-sphere approach, which involves assembling different numbers of spherical particles to mimic the shape of the original particle. This speeds up the computation and improves the accuracy in determining the particle–particle interactions [115]. The multi-sphere approach has been successfully applied to model various agricultural grains, including soybean [104, 113], corn [24], rice [116], wheat [65], peas, and beans [117]. However, particle behavior prediction accuracy depends on the number of sub-spheres used. The particles are interpreted as interlocked when the number of sub-spheres is small (≤ 9). Conversely, a higher number of sub-spheres results in a decrease in roughness [118]. To simulate complex-shaped particles within DEM, innovative methods have been developed that utilize three-dimensional shapes, including ellipsoidal [119–121], super-ellipsoid [122–124], sphero-cylinders and capsules [25, 125, 126], super-quadric [127–129], bonded particles [130–133], polyhedral [134, 135], and polygonal [136–138]. These advanced techniques require precisely determining key parameters such as contact points and normal and tangential contact vectors [35, 139].

Table 3.1 summarizes the particle shape models developed for predicting the behavior of the most common agricultural cereal grains. As presented, the multi-sphere approach is commonly used where the best number of spheres to represent a single grain or a seed kernel has not been confirmed yet. This could be because the complicated shape of agricultural grains and the presence of the surface crease make it challenging to develop particles with identical shapes. So, the number of spheres should be approximated by trials, depending on the computation time and prediction accuracy required.

A new large-scale DEM simulation model, namely, CG DEM modeling, which was subsequently proposed by Sakai's group [160–162], has been recently developed to overcome the problem of the high computational cost of a standard DEM. The parcels, or coarse-graining concept, are used at grain scale with a Lagrangian method, which has a governing equations algorithm similar to DEM [163, 164]. The general idea is that particles with the same properties, such as size and density, could be grouped into one CG particle (also called parcels). All the group particles will have the same translational and rotational movement in the

Table 3.1 Biogenic particle models used for DEM simulation

Particle type	DEM model/ reference	Particle shape	Research representations and investigations
Soybeans	2D single-circular disc model [140]	Circular disc	Simulating the separation of soybeans and mustard seeds using a sieve device under the condition of feeding rate and bed depth
	3D 4-sphere model [141]	Ellipsoid	Simulating the behavior of a single soybean kernel bouncing on aluminum, glass, and acrylic surfaces to measure the coefficient of restitution with high accuracy
	3D 4-sphere model [113]	Ellipsoid	Predicting the dry flow of soybean in a chute device compared with experimental data
	3D 4-sphere model [142]	Ellipsoid	Simulating soybeans flowing down on an inclined chute with a bumpy bottom to predict the soybean's flow velocity and force statistics during flow
	Comparison of 3D 1-sphere, 2-sphere, 3-sphere, and 4-sphere model [104]	Sphere and ellipsoid	Developing a particle model with appropriate parameters of coefficients of restitution, static friction, rolling friction, particle size distribution, and particle shear modulus that best matches the property values available in the literature. The best model is a single-sphere particle model that perfectly simulates soybean kernels in the bulk property tests
	Comparison of 3D 5-sphere, 9-sphere, and 13-sphere model [143]	Ellipsoid	Developing a particle model to analyze the discharging process of a silo. The result of the 5-sphere model was the best for modeling the soybean seed particle, which verified the feasibility and validity of the modeling methods for soybean seeds
Rice	The 2D single circular model [144]	Circular element	Modeling paddy and brown rice kernels in the shaking separation process compared to an experiment on the same scale
	3D triaxial particle model [110]	Ellipsoid	Simulating the impact behavior against the grain loss sensor. The DEM results were like the experimental data, providing evidence for optimizing grain loss sensors to accurately identify the rice seed impact signals
	3D 11-sphere model [116]	Ellipsoid	Simulating the rice kernels' filling, discharge flow, and piling with rolling friction coefficients of 0 and 0.3. Considering rolling friction, the pile shape is like the actual pile. Without rolling friction, particle mobility was higher, resulting in a spread of particles rather than a pile
	3D 7-sphere model [145]	Ellipsoid	Predicting the impact behavior of rice kernels on the impact board of an inclined elevator head with high accuracy
	3D single-spheroid model [23]	Spheroid	Simulating the separation process of rice kernels and straws using a coupled DEM and CFD model in an air and screen cleaning device

(continued)

Table 3.1 (continued)

Particle type	DEM model/ reference	Particle shape	Research representations and investigations
Corn	Comparison of 3D 1-sphere, 2-sphere, 5-sphere, and 13-sphere model [146]	Sphere and cluster shape	Studying the influence of corn shape and the interaction effects of DEM parameters such as static and rolling friction coefficients. The developed models were validated and calibrated by comparing the flow rate and corn pile with experimental tests of commercial screw augers. It is shown that the 2-sphere model is the best DEM corn shape model
	2D 2-disc clump model [147]	Disc	Modeling the internal friction angle and stiffness of grains where the calibration process was validated by modeling silo discharge and bucket filling
	Bonded particle model (BPM) [26]	Spheres	Simulating and optimizing the air-seeds flow in the aerodynamic seed-metering device using CFD–DEM. The model is reliable as a tool for understanding the physical phenomenon of seed movement in the gas phase and can be applied successfully to analyze the sources of difference
	3D 4-sphere model [148]	Cluster autonomous shape	Simulating the gravitational field's influence on the corn seeds' bulk responses under three different loadings
	3D 6-sphere model [149]	Cluster autonomous shape	Studying the pressure distributions, bulk density distributions, and flow properties during the filling and emptying of silos.
	Comparison of 3D single-sphere, 4-sphere, 6-sphere, 8-sphere, and 12-sphere model [150]	Cluster autonomous shape	Developing the most accurate modeling shape considering the rolling friction effect by calibrating silo discharging. Depending on the required accuracy, the approach of six sub-spheres could be sufficient to represent the shape of the grain of maize

(continued)

Table 3.1 (continued)

Particle type	DEM model/ reference	Particle shape	Research representations and investigations
Rapeseed	3D single-sphere model [59]	Sphere	Simulating rapeseed and wheat motion using a coupled DEM and CFD model in seed feeding device at different air velocities. The model can perfectly predict the physical phenomenon of seed movement in the airflow field
	3D single-sphere model [151]	Sphere	Predicting the rapeseed flow through a horizontal orifice using different contact models. The result showed that contact models give the same behavior for slow particle flow as experimental results but needed some improvement, including dissipation for higher particle flow rates
	3D single-sphere model [152]	Sphere	Simulating the deformation under bulk compressive loading of rapeseeds that there was a little deviation in the initial particle positions between the experiment and simulation
	3D single-sphere model [153]	Sphere	Investigating the influence of the moisture content of rapeseeds on the physical properties of grain bedding during uniaxial compression testing. The simulations and experiments showed differences in the elasticity and the stress transmission at various grain moisture contents. At the same time, the mechanical response is significantly affected by the moisture content of kernels
	3D single-sphere model [154]	Sphere	Simulating rapeseeds' free fall and impact against a flat surface using different contact models and moisture content
	2D single-circular disc model [155]	Circular disc	Modeling the bulk behavior of rapeseed during a direct shear test to determine the influence of three different levels of interparticle friction coefficient

(continued)

Table 3.1 (continued)

Particle type	DEM model/ reference	Particle shape	Research representations and investigations
Wheat	3D 3-sphere model [64]	Cluster shape	Predicting the velocity distribution of wheat kernels in a mixed-flow dryer where the obtained velocity was nearly accurate to the real
	Comparison of 3D single-sphere, 4-sphere, and 8-sphere model [156]	Cluster autonomous shape	Comparing the performance of different particle models in predicting the angle of internal friction and cohesion of wheat using linear and nonlinear contact models. The single-sphere particle model behaved better in the simulations than the multi-sphere models. Also, 3D particle models have higher accuracy in predicting the bulk behavior of wheat than a 2D approach
	2D single-circular disc model [157, 158]	Circular disc	Investigating a mixed-flow dryer's wheat velocity, moisture content, and residence time distributions. The DEM can adequately predict the main features of particle flow compared to the experimental results
	2D 5-disc ellipsoidal clump model [65]	Ellipsoid	Evaluating the traditional designs of mixed-flow dryers with different air ducts by coupling CFD–DEM to discover design deficits. The model can predict the actual flow pattern but not produce the expected dynamic angle of repose that typically forms under the air ducts
	Comparison of 3D single-sphere and multi-sphere model [159]	Sphere and kernel shape	Modeling the wheat in the first break milling process at different moisture content. The sphere-shaped model had a significant deviation from the experimental milling results. However, the two-shaped models could predict the trend observed for change in particle size of the first break stream with moisture content
Others	Bagasse particles	Rounded cylinder shape	Investigating theoretically the bagasse particle separation process in a pneumatic cylinder device using CFD–DEM to optimize the machine used for that purpose

considered system. In other words, many real particles of the same properties are represented by a small number of parcel/CG particles. One CG particle cannot contain particles of different sizes, as illustrated in Fig. 3.17 [165]. Therefore, the coarse grain scaling factor (f_{CG}) is defined as the size ratio of the CG particle (parcel) to the original particle ($d_{p_{CG}} = f_{CG} d_p$) and the number of particles in the coarse grain parcel is defined by ($N_{p_{CG}} = \frac{N_p}{f_{CG}^3}$) [166]. With this scaling up, the time step used for the

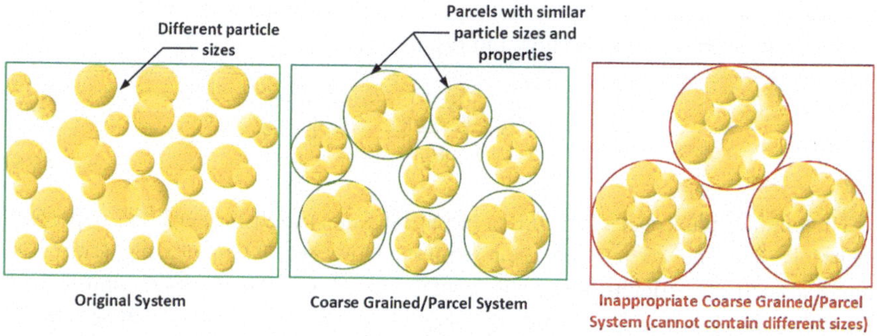

Fig. 3.17 Schematic representation of the CG DEM model concept

simulation also increases, leading to much faster computations and guaranteeing numerical accuracy by resolving interparticle collisions [90, 92, 167].

As for the existing DEM, the contact force acting on the CG particle is evaluated using springs, dash-pots, and a friction slider [161]. During the CG DEM modeling, some momentum and drag force corrections should be considered. In contrast, others, such as turbulence dispersion force, pressure gradient force, and virtual mass force, can be kept without modifications. For kinetic energy conservation, the velocity of the coarse grain system should be equal to that of the original particles [168]. To satisfy the pressure drop in the system, the fluid forces over the parcel should be equal to the sum of the fluid forces on each one of the original particles, given by ($\Delta P^{CG} = f_{CG}^3 \Delta P$) [165]. In addition, to make sure that both the CG and original systems can reach similar packing porosity, the mass m_p^{CG}, the volume vol_p^{CG}, and the gas–particle force of the parcel should be given by ($m_p^{CG} = f_{CG}^3 m_p$, $\mathrm{vol}_p^{CG} = f_{CG}^3 \mathrm{vol}_p$, $F_{gp}^{CG} = f_{CG}^3 F_{gp}$) [90, 161]. According to the assumption of the translational motion of the CG modeling, the drag force acting on a CG particle becomes f_{CG}^3 times larger than that of the original particle, as given in Eq. (3.4) [161].

$$F_{\mathrm{drag}_{CG}} = \frac{1}{2} f_{CG}^3 C_D \rho_g A_{CG}' \left|U_g - U_p\right| \left(U_g - U_p\right) \tag{3.4}$$

where A_{CG}' is the area of the parcel, which is equal to ($f_{CG}^2 A'$), A' is the actual particle area, and f_{CG}^3 is the particle drag correlation. The drag coefficient, C_D, also needs to be corrected, as the drag coefficient calculated for the parcel should have the same value as that of a single original particle to scale the drag force correctly [169]. Since the drag coefficient is usually a function of the Reynolds number and the volume fraction of the particles, these two numbers must be kept the same for the original system to obtain the same drag coefficient. The volume of the particles is kept the same when scaling up, so there is no need to change the volume fraction calculation [165]. As the scaled particle size $d_{p_{CG}}$ is used for the parcel Reynolds number calculation, the gas viscosity, μ_g, is also scaled with f_{CG}, aiming to keep the same Reynolds number for the scaled parcel and the original particle. The corrected Reynolds number equation is shown in Eq. (3.5) [166].

$$\mathrm{Re}_{p_{CG}} = \frac{\rho_p d_{p_{CG}}}{f_{CG}\,\mu_g}\left|\left(\overrightarrow{u_p} - \overrightarrow{u_g}\right)\right| \tag{3.5}$$

The contact plane between spherical particles is straightforward and is perpendicular to the line connecting their centers. However, when dealing with non-spherical particles, determining the contact plane can be a more complicated. Depending on the particle's shape and orientation, the contact plane can be perpendicular to the line connecting the closest points of the two particles, the line connecting the nearest points of a particle and a triangle, or the line connecting two points with the maximum overlap distance [22, 139, 170]. To accurately model and simulate these complex, irregularly shaped particles, we must treat them as polyhedrons, three-dimensional bodies with many corners, edges, and faces. Each polyhedron is subdivided into tetrahedron elements, which are bonded to neighboring elements through the triangular faces, as depicted in Fig. 3.18 [171–175]. As these elements move, their relative motion generates linear and angular deformations on the hypothetical bonds, producing forces and moments that resist normal, tangential, bending, and torsional deformations. To model these forces and moments in a DEM, the elastic and damping forces are summed relating to each hypothetical bond, as indicated by Eqs. (3.6) and (3.7). These sophisticated models and methods can provide a more accurate and comprehensive understanding of the complex behavior and interactions of non-spherical particles in various environments.

$$F_b^n = K_b^n X_{\mathrm{rel}}^n + C_b^n v_{\mathrm{rel}}^n \tag{3.6}$$

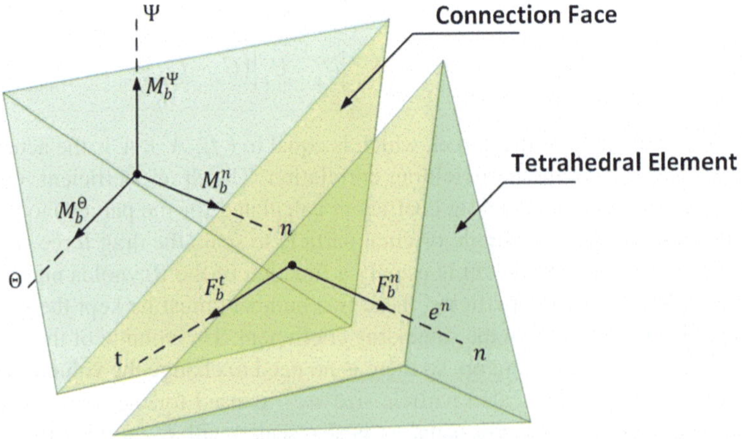

Fig. 3.18 Forces and moments acting on a tetrahedral element in a polyhedral particle

$$F_b^t = K_b^t X_{\text{rel}}^t + C_b^t v_{\text{rel}}^t \tag{3.7}$$

where F_b^n and F_b^t are the normal and tangential (elastic and damping) forces, respectively; K_b^n and K_b^t are the normal and tangential stiffnesses, respectively; C_b^n and C_b^t are the normal and tangential damping coefficients, respectively; X_{rel}^n and X_{rel}^t are the normal and tangential relative displacements between the center points of the two elements connected by the hypothetical bond, respectively; and v_{rel}^n and v_{rel}^t are the normal and tangential relative velocities between the center points of the two elements connected by the hypothetical bond, respectively.

The normal and tangential relative displacement components can be calculated according to Eqs. (3.8) and (3.9):

$$X_{\text{rel}}^n = X_{\text{rel}} e^n \tag{3.8}$$

$$X_{\text{rel}}^t = X_{\text{rel}} - X_{\text{rel}}^n \tag{3.9}$$

where e^n is defined by the unit vector in the normal direction. In brief, the relative velocity component can be determined and analyzed similarly to the relative displacement. Furthermore, the values of the normal and tangential stiffnesses for the hypothetical bond can be determined by Eqs. (3.10) and (3.11) [176]:

$$K_b^n = \frac{E_R E_i A}{\sigma l_t} \tag{3.10}$$

$$K_b^t = K_r K_b^n \tag{3.11}$$

where E_R is the ratio between the normal and tangential Young's modulus [177, 178], A is the area of the triangular contacting face, σ is the dimensionless Young's modulus and can be defined by Eq. (3.12), l_t is the average length of the edges of all tetrahedra in the particle, and K_r is the stiffness ratio determined by Eq. (3.13).

$$\sigma = \frac{E_i}{K_{na}^n l_t} \tag{3.12}$$

$$K_r = \frac{K_i^t}{K_i^n} \tag{3.13}$$

where K_{na}^n is the normal stiffness per unit area, and K_i^t and K_i^n are the tangential and normal stiffness of the particle, respectively. On the other hand, the three components of the momentum are calculated according to Eqs. (3.14), (3.15), and (3.16) [171]:

$$M_b^n = \frac{K_b^n I_n}{A}\theta_{\text{rel}}^n + \frac{C_b^n I_n}{A}\omega_{\text{rel}}^n \tag{3.14}$$

$$M_b^\Psi = \frac{K_b^t I_\Psi}{A}\theta_{\text{rel}}^\Psi + \frac{C_b^t I_\Psi}{A}\omega_{\text{rel}}^\Psi \tag{3.15}$$

$$M_b^\Theta = \frac{K_b^t I_\Theta}{A}\theta_{\text{rel}}^\Theta + \frac{C_b^n I_\Theta}{A}\omega_{\text{rel}}^\Theta \tag{3.16}$$

where M_b^n, M_b^Ψ, and M_b^Θ are the momentum components related to the hypothetical bond around axes corresponding to the orthogonal directions n, Ψ, and Θ, respectively; θ_{rel}^n, θ_{rel}^Ψ, and $\theta_{\text{rel}}^\Theta$ are the relative angular displacements between the bonded elements; ω_{rel}^n, ω_{rel}^Ψ, and $\omega_{\text{rel}}^\Theta$ are the relative angular velocities between the bonded elements; and I_n, I_Ψ, and I_Θ are moments of inertia of the rectangular face associated with the virtual bond.

Since the particle becomes longer and more pliable, such as straw, a more flexible approach is necessary. This is where the concept of multiple elements comes into play [179]. By connecting sphero-cylinders through virtual bonds, a flexible fiber particle can be created. Figure 3.19 presents the forces and moments acting on a long fiber particle and a custom multibranched fiber with several coincident joints. In Figure 3.19a, a flexible straight fiber is segmented into sphero-cylindrical elements, while Fig. 3.19b illustrates a custom fiber with six coincident joints at a specific point. Fibers are primarily one-dimensional shapes that can be described by a line or group of lines in three-dimensional space, as depicted in Fig. 3.19a. The central line of a straight fiber is divided into segments that generate sphero-cylindric elements, allowing for the same number of segments as the number of elements. The endpoints of a segment align with the center points that generate the hemispherical cap surfaces

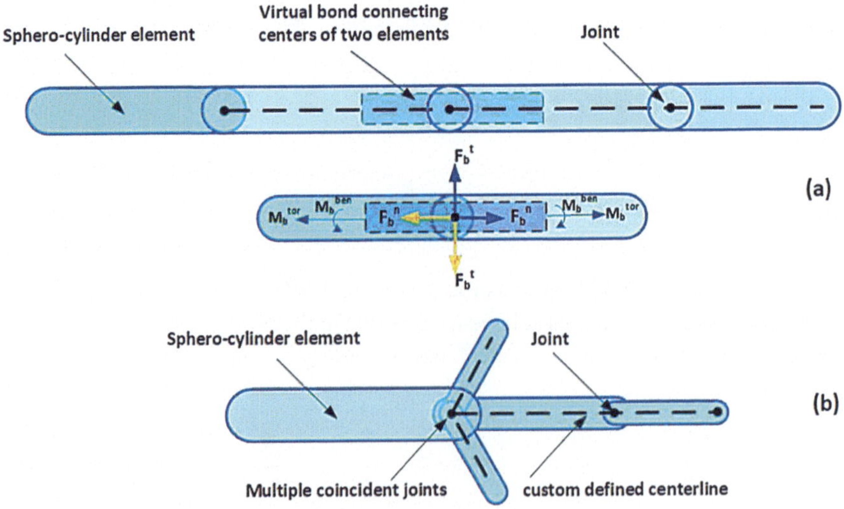

Fig. 3.19 Schematic representation of a flexible fiber particle

at both ends of an element. Additionally, in a fiber's initial, undeformed state, the endpoints of two adjacent elements always coincide. When modeling the dynamics of a flexible fiber, opposing forces and moments that resist deformations are considered to act at the endpoints of the centerline segments of two adjacent sphero-cylinders as shown in Fig. 3.19b. These forces and moments are believed to be the reaction to deformations of a joint entity that connects those points. Therefore, utilizing multiple elements and virtual bonds allows for a more adaptable and flexible approach when dealing with longer, more pliable particles such as fibers.

A virtual bond is an auxiliary entity possessing both elastic and viscous properties. It joins the centers of two adjacent sphero-cylinder elements, which initially coincide. As these elements move, the virtual bonds may undergo linear and angular deformations, prompting the generation of forces and moments. These forces and moments resist normal, tangential, bending, and torsional deformations. Each force and moment linked to a virtual bond is modeled as the sum of two components, namely, elastic and damping, where the normal and shear forces can be calculated as expressed in Eqs. (3.3) and (3.4). The bending and torsional moments acting at the ends of the virtual bonds are expressed in Eqs. (3.17) and (3.18).

$$M_b^{\text{ben}} = \frac{K_b^n I}{A}\theta^{\text{ben}} + \frac{C_b^n I}{A}\omega_{\text{rel}}^{\text{ben}} \tag{3.17}$$

$$M_b^{\text{tor}} = \frac{K_b^t J}{A}\theta^{\text{tor}} + \frac{C_b^t I}{A}\omega_{\text{rel}}^{\text{tor}} \tag{3.18}$$

where I and J are the area moment of inertia and the polar area moment of inertia, respectively, and of the circular cross-section of the virtual bonds, θ^{ben} and θ^{tor} are the bending and twisting angular displacements between the two adjacent sphero-cylinder elements, respectively.

In addition to the conventional straight fiber shapes, fiber manipulation extends to intricate, multi-branched geometries. Dubbed "custom fibers," these structures can be defined by providing appropriate data via an auxiliary file. The minimum data required for constructing a custom fiber includes the coordinates of each segment's endpoints along the fiber's centerline, as well as the diameter of the corresponding sphero-cylinder element. The image depicted in Fig. 3.19b showcases how this type of fiber can possess elements with varying diameters. Nevertheless, the center points of each hemispherical cap surface will consistently align with the endpoints of segments in the fiber's centerline geometry. When multiple centerline segments converge at the exact location, multi-branched geometries arise, where the endpoints of four segments coincide. To account for these intersections, a joint will connect each possible pair of segments, resulting in six joints coinciding at point in the example shown in Fig. 3.19b. These joints behave identically to those in the straight fibers, exerting forces and moments that resist changes in relative positions among the elements they connect. Moreover, depending on the total number of elements specified for the flexible custom fiber, each segment within the provided centerline geometry can still be subdivided into smaller segments. This flexibility enables a custom fiber to be viewed as a conglomeration of numerous flexible straight fibers. Therefore, the possibilities for creating bespoke, multi-branched fibers are virtually limitless, thanks to the versatility and adaptability of modern fiber manipulation techniques.

3.9 Summary

Aerodynamic systems in agriculture, including grain dryers, spouted beds, aerodynamic separators, and combine harvesters, have seen significant advancements in both performance and design. This progress has been made possible by integrating numerical models such as DEM and CFD. DEM, known for its ability to simulate the movements of discrete particles under various forces, has emerged as a powerful tool for understanding dense phase flows in agricultural processing. The coupling of CFD with DEM has played a crucial role in overcoming the limitations of CFD multiphase models, particularly in estimating particulate matter phases in gas–particle flows. Although this coupled approach increases computational time, it allows for a detailed analysis of particle–particle interactions and interactions between particles and the surrounding gas or fluid. Researchers have meticulously studied and modeled the geometry components and parameters of aerodynamic systems, leading to significant advancements in the efficiency of processes such as separation, cleaning, sorting, grading, and density classification.

Looking ahead, the continued integration of advanced mathematical, numerical, and computational techniques promises further innovations in aerodynamics for agriculture. With advancing technology and improving modeling capabilities, we can expect more efficient and effective solutions for optimizing aerodynamic systems in the agricultural industry. Embracing these advancements will undoubtedly drive success, sustainability, and productivity in agricultural processes for years to come.

References

1. Piven VV (2018) Determination of the extent of fraction in air separation of grain material. J Phys Conf Ser 1059. https://doi.org/10.1088/1742-6596/1059/1/012001
2. Clarke B (1985) Cleaning seeds by fluidization. J Agric Eng Res 31:231–242. https://doi.org/10.1016/0021-8634(85)90090-3
3. Gorial BY, O'Callaghan JR (1990) Aerodynamic properties of grain/straw materials. J Agric Eng Res 46:275–290. https://doi.org/10.1016/S0021-8634(05)80132-5
4. Zewdu AD (2007) Aerodynamic properties of tef grain and straw material. Biosyst Eng 98:304–309. https://doi.org/10.1016/j.biosystemseng.2007.08.003
5. Samantaray SK, Mohapatra SS, Munshi B (2018) Experimental findings and analysis of terminal velocity and drag coefficient of Raschig ring in vertical and inclined channel. Powder Technol 340:440–448. https://doi.org/10.1016/j.powtec.2018.09.030
6. Zhang L, Honaker R, Liu W et al (2015) Calculation of terminal velocity in transitional flow for spherical particle. Int J Min Sci Technol 25:311–317. https://doi.org/10.1016/j.ijmst.2015.02.022

7. Czernik Z (1983) Studies of aerodynamical features of seeds of scotch pine, Norway spruce and European larch (Badania wlasciwosci aerodynamicznych nasion sosny zwyczajnej, swierka pospolitego i modrzewia europejskieg). Sylwan 9:31–40
8. Tylek P, Walczyk J (2003) Critical air velocity as a separation feature in nuts of european beech (Fagus sylvatica L.). Electron J Polish Agric Univ: Univ 6(2):4
9. Ghamari S, Borghei AM, Rabbani H et al (2010) Modeling the terminal velocity of agricultural seeds with artificial neural networks. Afr J Agric Res 5:389–398. https://doi.org/10.5897/AJAR09.626
10. Khodabakhshian R, Emadi B, Khojastehpour M, Golzarian MR (2018) Aerodynamic separation and cleaning of pomegranate arils from rind and white segments (locular septa). J Saudi Soc Agric Sci 17:61–68. https://doi.org/10.1016/j.jssas.2016.01.003
11. Mohsenin NN (1986) Physical properties of plant and animal materials. Gordon and Breach Science, New York
12. Tylek P, Walczyk J (2004) Effectiveness of the pneumatic separation of Norway spruce Picea abies (L.) karst seeds. Dendrobiology 51:101–104
13. Levenspiel O, Haider A (1989) Drag coefficient and terminal velocity of spherical and nonspherical particles. Powder Technol 58:63–70. https://doi.org/10.1016/0032-59109(89)80008-7
14. De Almeida E, Spogis N, Taranto OP, Silva MA (2019) Theoretical study of pneumatic separation of sugarcane bagasse particles. Biomass Bioenergy 127:105256. https://doi.org/10.1016/j.biombioe.2019.105256
15. Lapple C (1951) Fluid and particle mechanics, 1st edn. University of Delaware, Newark
16. Hauhouot-O'Hara M, Criner BR, Brusewitz GH, Solie JB (2000) Selected physical characteristics and aerodynamic properties of cheat SEED for separation from wheat. J Sci Res Dev 2:1–14
17. Afonso Júnior PC, Corrêa PC, Pinto FAC, Queiroz DM (2007) Aerodynamic properties of coffee cherries and beans. Biosyst Eng 98:39–46. https://doi.org/10.1016/j.biosystemseng.2007.03.010
18. Shellard JE, Macmillan RH (1978) Aerodynamic properties of threshed wheat materials. J Agric Eng Res 23:273–281. https://doi.org/10.1016/0021-8634(78)90101-4
19. Shahbazi F (2013) Aerodynamic properties of wild mustard (Sinapis arvensis L.) seed for separation from canola. J Sci Food Agric 93:1466–1470. https://doi.org/10.1002/jsfa.5916
20. Binelo MO, de Lima RF, Khatchatourian OA, Stránský J (2019) Modelling of the drag force of agricultural seeds applied to the discrete element method. Biosyst Eng 178:168–175. https://doi.org/10.1016/j.biosystemseng.2018.11.013
21. Khatchatourian OA, Binelo MO, de Lima RF, De Lima RF (2014) Simulation of soya bean flow in mixed-flow dryers using DEM. Biosyst Eng 123:68–76. https://doi.org/10.1016/j.biosystemseng.2014.05.003
22. Guo Y, Curtis JS (2015) Discrete element method simulations for complex granular flows. Annu Rev Fluid Mech 47:21–46. https://doi.org/10.1146/annurev-fluid-010814-014644
23. Li H, Li Y, Gao F et al (2012) CFD-DEM simulation of material motion in air-and-screen cleaning device. Comput Electron Agric 88:111–119. https://doi.org/10.1016/j.compag.2012.07.006
24. Ren B, Zhong W, Chen Y et al (2012) CFD-DEM simulation of spouting of corn-shaped particles. Particuology 10:562–572. https://doi.org/10.1016/j.partic.2012.03.011
25. Liu X, Gan J, Zhong W, Yu A (2019) Particle shape effects on dynamic behaviors in a spouted bed: CFD-DEM study. Powder Technol. https://doi.org/10.1016/j.powtec.2019.07.099
26. Han D, Zhang D, Jing H et al (2018) DEM-CFD coupling simulation and optimization of an inside-filling air-blowing maize precision seed-metering device. Comput Electron Agric 150:426–438. https://doi.org/10.1016/j.compag.2018.05.006
27. Tsuji Y (2007) Multi-scale modeling of dense phase gas–particle flow. Chem Eng Sci 62:3410–3418. https://doi.org/10.1016/j.ces.2006.12.090
28. Tsuji Y, Tanaka T, Ishida T (1992) Lagrangian numerical-simulation of plug flow of cohesionless particles in a horizontal pipe. Powder Technol 71:239–250. https://doi.org/10.1016/0032-5910(92)88030-L

29. Zhou ZY, Kuang SB, Chu KW, Yu AB (2010) Discrete particle simulation of particle–fluid flow: model formulations and their applicability. J Fluid Mech 661:482–510. https://doi.org/10.1017/S002211201000306X
30. Wang B, Chu KW, Yu AB (2007) Numerical study of particle–fluid flow in a hydrocyclone. Ind Eng Chem Res 46:4695–4705. https://doi.org/10.1021/ie061625u
31. Xu BH, Yu AB (1997) Numerical simulation of the gas–solid flow in a fluidized bed by combining discrete particle method with computational fluid dynamics. Chem Eng Sci 52:2785–2809. https://doi.org/10.1016/S0009-2509(97)00081-X
32. Chu KW, Wang B, Yu AB, Vince A (2009) CFD–DEM modelling of multiphase flow in dense medium cyclones. Powder Technol 193:235–247. https://doi.org/10.1016/j.powtec.2009.03.015
33. Cortés C, Gil A (2007) Modeling the gas and particle flow inside cyclone separators. Prog Energy Combust Sci 33:409–452. https://doi.org/10.1016/j.pecs.2007.02.001
34. Derksen JJ, Sundaresan S, van den Akker HEA (2006) Simulation of mass-loading effects in gas-solid cyclone separators. Powder Technol 163:59–68. https://doi.org/10.1016/B978-008044544-1/50089-3
35. Horabik J, Molenda M (2016) Parameters and contact models for DEM simulations of agricultural granular materials: a review. Biosyst Eng 147:206–225. https://doi.org/10.1016/j.biosystemseng.2016.02.017
36. Almeida E, Spogis N, Silva MA (2016) Computational study of the pneumatic separation of sugarcane bagasse. In: 6th international conference on engineering for waste and biomass valorisation
37. Fonte CB Jr, Oliveira JAA, De Almeida LC (2015) Dem-Cfd coupling: mathematical modelling and case studies using Rocky-Dem® and Ansys Fluent®. In: Eleventh international conference on CFD in the minerals and process industries. CSIRO, Melbourne
38. Chu KW, Wang B, Xu DL et al (2011) CFD-DEM simulation of the gas-solid flow in a cyclone separator. Chem Eng Sci 66:834–847. https://doi.org/10.1016/j.ces.2010.11.026
39. Romaní Fernández X, Nirschl H (2013) Simulation of particles and sediment behaviour in centrifugal field by coupling CFD and DEM. Chem Eng Sci 94:7–19. https://doi.org/10.1016/j.ces.2013.02.039
40. Wang Y, Williams K, Jones M, Chen B (2017) CFD simulation methodology for gas-solid flow in bypass pneumatic conveying—a review. Appl Therm Eng 125:185–208. https://doi.org/10.1016/j.applthermaleng.2017.05.063
41. Ullah A, Hong K, Gao Y et al (2019) An overview of Eulerian CFD modeling and simulation of non-spherical biomass particles. Renew Energy 141:1054–1066. https://doi.org/10.1016/j.renene.2019.04.074
42. Norouzi HR, Zarghami R, Sotudeh-gharebagh R, Mostoufi N (2016) Coupled CFD-DEM modeling: formulation, implementation and application to multiphase flows, 1st edn. Wiley, Newark
43. B. A. Adewumi (2008) 2D modeling of grain transport and separation in the horizontal air stream. In: Food processing automation conference CD-Rom, pp 1–10
44. Wang Q, Melaaen MC, De Silva SR (2001) Investigation and simulation of a cross-flow air classifier. Powder Technol 120:273–280. https://doi.org/10.1016/S0032-5910(01)00282-0
45. Guo Y, Wu CY, Kafui KD, Thornton C (2011) 3D DEM/CFD analysis of size-induced segregation during die filling. Powder Technol 206:177–188. https://doi.org/10.1016/j.powtec.2010.05.029
46. Badretdinov I, Mudarisov S, Lukmanov R et al (2019) Mathematical modeling and research of the work of the grain combine harvester cleaning system. Comput Electron Agric 165:104966. https://doi.org/10.1016/j.compag.2019.104966
47. Mudarisov S, Khasanov E, Rakhimov Z et al (2017) Specifying two-phase flow in modeling pneumatic systems performance of farm machines specifying two-phase flow in modeling pneumatic systems performance of farm machines. J Mech Eng Res Dev 40:706–715. https://doi.org/10.7508/jmerd.2017.04.018

48. Li Y, Xu L, Zhou Y et al (2018) Effects of throughput and operating parameters on cleaning performance in air-and-screen cleaning unit: a computational and experimental study. Comput Electron Agric 152:141–148. https://doi.org/10.1016/j.compag.2018.07.019
49. Lenaerts B, Aertsen T, Tijskens E et al (2014) Simulation of grain-straw separation by discrete element modeling with bendable straw particles. Comput Electron Agric 101:24–33. https://doi.org/10.1016/j.compag.2013.12.002
50. Ma Z, Li Y, Xu L (2015) Discrete-element method simulation of agricultural particles' motion in variable-amplitude screen box. Comput Electron Agric 118:92–99. https://doi.org/10.1016/j.compag.2015.08.030
51. Enchen J, Zhanfeng S, Zhiyang P, Lijun W (2014) Numerical simulation based on CFD-DEM and experiment of grain moving laws in inertia separation chamber. J Agric Mach 45:117–122. https://doi.org/10.6041/j.issn.1000-1298
52. Yuan J, Wu C, Li H et al (2018) Movement rules and screening characteristics of rice-threshed mixture separation through a cylinder sieve. Comput Electron Agric 154:320–329. https://doi.org/10.1016/j.compag.2018.09.012
53. Balafoutis A, Beck B, Fountas S et al (2017) Precision agriculture technologies positively contributing to GHG emissions sustainability precision agriculture technologies positively contributing to GHG emissions mitigation. Farm Prod Econ Sustain 9:1339. https://doi.org/10.3390/su9081339
54. Minfeng J, Yongqian D, Hongfeng Y et al (2018) Optimal structure design and performance tests of seed metering device with fluted rollers for precision wheat seeding machine. IFAC Pap OnLine 17:509–514. https://doi.org/10.1016/j.ifacol.2018.08.158
55. Singh RC, Singh G, Saraswat DC (2005) Optimisation of design and operational parameters of a pneumatic seed metering device for planting cottonseeds. Biosyst Eng 92:429–438. https://doi.org/10.1016/j.biosystemseng.2005.07.002
56. Jianbo Z, Junfang X, Yong Z, Shun Z (2014) Design and experimental study of the control system for precision seed-metering device. Int J Agric Biol Eng 7:13–18. https://doi.org/10.3965/j.ijabe.20140703.002
57. Kamgar S, Eslami M, Maharlouie M (2013) Design, development and evaluation of a mechatronic transmission system to improve the performance of a conventional row crop planter. Intl J Agron Plant Prod 4:480–487
58. Dandan H, Dongxing Z, Li Y et al (2017) EDEM-CFD simulation and experiment of working performance of inside-filling air-blowing seed metering device in maize. Trans Chin Soc Agric Eng 33:23–31
59. Lei X, Liao Y, Liao Q (2016) Simulation of seed motion in seed feeding device with DEM-CFD coupling approach for rapeseed and wheat. Comput Electron Agric 131:29–39. https://doi.org/10.1016/j.compag.2016.11.006
60. Lei X, Liao Y, Zhang Q et al (2018) Numerical simulation of seed motion characteristics of distribution head for rapeseed and wheat. Comput Electron Agric 150:98–109. https://doi.org/10.1016/j.compag.2018.04.009
61. Weigler F, Mellmann J (2014) Investigation of grain mass flow in a mixed flow dryer. Particuology 12:33–39. https://doi.org/10.1016/j.partic.2013.04.004
62. Bacelos MS, Almeida PIF (2011) Modelling of drying kinetic of potatoes taking into account shrinkage. Procedia Food Sci 1:713–721. https://doi.org/10.1016/j.profoo.2011.09.108
63. Oksanen T (2018) Controlling air flow in recirculating mixed flow batch dryer with double bed mode. Comput Electron Agric 149:133–138. https://doi.org/10.1016/j.compag.2017.07.018
64. Keppler I, Kocsis L, Oldal I et al (2012) Grain velocity distribution in a mixed flow dryer. Adv Powder Technol 23:824–832. https://doi.org/10.1016/j.apt.2011.11.003
65. Weigler F, Scaar H, Mellmann J (2012) Investigation of particle and air flows in a mixed-flow dryer. Dry Technol 30:1730–1741. https://doi.org/10.1080/07373937.2012.703742
66. Weigler F, Mellmann J, Franke G, Scaar H (2013) Experimental studies on a newly developed mixed-flow dryer. Dry Technol 31:1736–1743. https://doi.org/10.1080/07373937.2013.806929

67. Scaar H, Franke G, Weigler F et al (2016) Experimental and numerical study of the airflow distribution in mixed-flow grain dryers. Dry Technol 34:595–607. https://doi.org/10.1080/07373937.2015.1064946
68. Azmir J, Hou Q, Yu A (2019) CFD-DEM simulation of drying of food grains with particle shrinkage. Powder Technol 343:792–802. https://doi.org/10.1016/j.powtec.2018.11.097
69. Azmir J, Hou Q, Yu A (2019) CFD-DEM study of the effects of food grain properties on drying and shrinkage in a fluidised bed. Powder Technol. https://doi.org/10.1016/j.powtec.2019.10.021
70. Yuu S, Jotaki T, Tomita Y, Yoshida K (1978) The reduction of pressure drop due to dust loading in a conventional cyclone. Chem Eng Sci 33:1573–1580. https://doi.org/10.1016/0009-2509(78)85132-X
71. Hoffmann AC, de Jonge R, Arends H, Hanrats C (1995) Evidence of the "natural vortex length" and its effect on the separation efficiency of gas cyclones. Filtr Sep 32:799–804. https://doi.org/10.1016/S0015-1882(97)84131-6
72. Faulkner WB, Buser MD, Whitelock DP, Shaw BW (2007) Effects of cyclone diameter on performance of 1D3D cyclones: collection efficiency. Trans ASABE 50:1053–1060
73. Utikar R, Darmawan N, Tade M et al (2010) Hydrodynamic simulation of cyclone separators. In: OH HW (ed) Computational fluid dynamics. INTECH, Croatia, pp 241–263
74. Boysan F, Ayers WH, Swithenbank J (1982) A fundamental mathematical modeling approach to cyclone design. Trans Inst Chem Eng 60:222–230
75. Slack MD, Prasad RO, Bakker A, Boysan F (2000) Advances in cyclone modelling using unstructured grids. Chem Eng Res Des 78:1098–1104. https://doi.org/10.1205/026387600528373
76. Wang S, Luo K, Hu C, Fan J (2017) CFD-DEM study of the effect of cyclone arrangements on the gas-solid flow dynamics in the full-loop circulating fluidized bed. Chem Eng Sci 172:199–215. https://doi.org/10.1016/j.ces.2017.05.052
77. Qian F, Huang Z, Chen G, Zhang M (2007) Numerical study of the separation characteristics in a cyclone of different inlet particle concentrations. Comput Chem Eng 31:1111–1122. https://doi.org/10.1016/j.compchemeng.2006.09.012
78. Xiang RB, Lee KW (2005) Numerical study of flow field in cyclones of different height. Chem Eng Process 44:877–883. https://doi.org/10.1016/j.cep.2004.09.006
79. Wang B, Xu DL, Chu KW, Yu AB (2006) Numerical study of gas-solid flow in a cyclone separator. Appl Math Model 30:1326–1342. https://doi.org/10.1016/j.apm.2006.03.011
80. Narasimha M, Brennan MS, Holtham PN (2006) Numerical simulation of magnetite segregation in a dense medium cyclone. Miner Eng 19:1034–1047. https://doi.org/10.1016/j.mineng.2006.03.013
81. Bernardo S, Mori M, Peres AP, Dionı RP (2006) 3-D computational fluid dynamics for gas and gas-particle flows in a cyclone with different inlet section angles. Powder Technol 162:190–200. https://doi.org/10.1016/j.powtec.2005.11.007
82. Zhao B, Su Y, Zhang J (2006) Simulation of gas flow pattern and separation efficiency in cyclone with conventional single and spiral double inlet configuration. Chem Eng Res Des 84:1158–1165. https://doi.org/10.1205/cherd06040
83. De Brito Dias D, Mori M, Martignoni WP (2008) Study of different approaches for modeling cyclones using CFD. In: AIChE annual meeting
84. Meier H, Vegini A, Mori M (2011) Four-phase eulerian-eulerian model for prediction of multiphase flow in cyclones. J Comput Multiph Flows 3:93–105. https://doi.org/10.1260/1757-482X.3.2.93
85. Gimbun J (2008) CFD simulation of aerocyclone hydrodynamics and performance at extreme temperature. Eng Appl Comput Fluid Mech 2:22–29. https://doi.org/10.1080/19942060.2008.11015208
86. Shukla SK, Shukla P, Ghosh P (2013) The effect of modeling of velocity fluctuations on prediction of collection efficiency of cyclone separators. Appl Math Model 37:5774–5789. https://doi.org/10.1016/j.apm.2012.11.019

87. Elsayed K (2015) Optimization of the cyclone separator geometry for minimum pressure drop using co-kriging. Powder Technol 269:409–424. https://doi.org/10.1016/j.powtec.2014.09.038
88. Kumar A, Roy AK (2015) Numerical simulation of cyclone separators with different main body diameter. Int J Appl Eng Res 10:4225–4230
89. Song C, Pei B, Jiang M et al (2016) Numerical analysis of forces exerted on particles in cyclone separators. Powder Technol 294:437–448. https://doi.org/10.1016/j.powtec.2016.02.052
90. Chu K, Chen J, Yu A (2016) Applicability of a coarse-grained CFD–DEM model on dense medium cyclone. Miner Eng 90:43–54. https://doi.org/10.1016/j.mineng.2016.01.020
91. Chu K, Chen Y, Ji L et al (2022) Coarse-grained CFD-DEM study of gas-solid flow in gas cyclone. Chem Eng Sci 260:117906. https://doi.org/10.1016/j.ces.2022.117906
92. Napolitano ES, Di Renzo A, Di Maio FP (2022) Coarse-grain DEM-CFD modelling of dense particle flow in gas–solid cyclone. Sep Purif Technol 287:120591. doi:https://doi.org/10.1016/j.seppur.2022.120591
93. Kozołub P, Klimanek A, Białecki RA, Adamczyk WP (2017) Numerical simulation of a dense solid particle flow inside a cyclone separator using the hybrid Euler–Lagrange approach. Particuology 31:170–180. https://doi.org/10.1016/j.partic.2016.09.003
94. Zhang ZW, Li Q, Zhang YH, Wang HL (2022) Simulation and experimental study of effect of vortex finder structural parameters on cyclone separator performance. Sep Purif Technol 286:120394. https://doi.org/10.1016/j.seppur.2021.120394
95. Wargadalam VJ, Andira MA, Putra CA et al (2019) Numerical simulation of a small-scale cyclone separator using MP-PIC method. IOP Conf Ser Mater Sci Eng 694:012015
96. Zhou H, Hu Z, Zhang Q et al (2019) Numerical study on gas-solid flow characteristics of ultra-light particles in a cyclone separator. Powder Technol 344:784–796. https://doi.org/10.1016/j.powtec.2018.12.054
97. Wang S, Li H, Wang R et al (2019) Effect of the inlet angle on the performance of a cyclone separator using CFD-DEM. Adv Powder Technol 30:227–239. https://doi.org/10.1016/j.apt.2018.10.027
98. Nemati J, Beheshti B, Borghei AM (2020) Cfd-based numerical simulation of cyclone separator for separating stigmas from petals of saffron. J Agric Eng 51:239–247. https://doi.org/10.4081/jae.2020.1091
99. Chlebnikovas A, Kilikevičius A, Selech J et al (2021) The numerical modeling of gas movement in a single inlet new generation multi-channel cyclone separator. Energies 14. https://doi.org/10.3390/en14238092
100. Mirzaei M, Jensen PA, Nakhaei M et al (2022) A hybrid multiphase model accounting for particle agglomeration for coarse-grid simulation of dense solid flow inside large-scale cyclones. Powder Technol 399:117186. https://doi.org/10.1016/j.powtec.2022.117186
101. Morin M, Raynal L, Karri SBR, Cocco R (2021) Effect of solid loading and inlet aspect ratio on cyclone efficiency and pressure drop: experimental study and CFD simulations. Powder Technol 377:174–185. https://doi.org/10.1016/j.powtec.2020.08.052
102. Wan G, Sun G, Xue X, Shi M (2008) Solids concentration simulation of different size particles in a cyclone separator. Powder Technol 183:94–104. https://doi.org/10.1016/j.powtec.2007.11.019
103. Duran J (2000) Sands, powders, and grains: an introduction to the physics of granular materials. Springer, New York
104. Boac JM, Casada ME, Maghirang RG (2010) Material and interaction properties of selected grains and oilseeds for modeling discrete particles. Trans ASABE 53:1201–1216. https://doi.org/10.13031/2013.32577
105. Favier JF, Abbaspour-Fard MH, Kremmer M, Raji AO (1999) Shape representation of axi-symmetrical, non-spherical particles in discrete element simulation using multi-element model particles. Eng Comput 16:467–480. https://doi.org/10.1108/02644409910271894
106. Hogu C (1998) Shape representation and contact detection for discrete element simulations of arbitrary geometries. Eng Comput 15:374–390

107. Zhong W, Yu A, Liu X et al (2016) DEM/CFD-DEM modelling of non-spherical particulate systems: theoretical developments and applications. Powder Technol 302:108–152. https://doi.org/10.1016/j.powtec.2016.07.010
108. Kureck H, Govender N, Siegmann E et al (2019) Industrial scale simulations of tablet coating using GPU based DEM: a validation study. Chem Eng Sci 202:462–480. https://doi.org/10.1016/j.ces.2019.03.029
109. ISO 3435 (1977) Continuous mechanical handling equipment classification and symbolization of bulk materials, 1st edn. International Organization for Standardization, Geneva
110. Zhan Z, Yaoming L, Zhenwei L, Zhiqiang G (2013) DEM simulation and physical testing of rice seed impact against a grain loss sensor. Biosyst Eng 116:410–419. https://doi.org/10.1016/j.biosystemseng.2013.10.002
111. Athanassiadis AG, Miskin MZ, Kaplan P et al (2014) Particle shape effects on the stress response of granular packings. Soft Matter 10:48–59. https://doi.org/10.1039/c3sm52047a
112. Boac JM, Ambrose RPK, Casada ME et al (2014) Applications of discrete element method in modeling of grain postharvest operations. Food Eng Rev 6:128–149. https://doi.org/10.1007/s12393-014-9090-y
113. Vu-Quoc L, Zhang X, Walton OR (2000) A 3-D discrete-element method for dry granular flows of ellipsoidal particles. Comput Methods Appl Mech Eng 187:483–528. https://doi.org/10.1016/S0045-7825(99)00337-0
114. Kruggel-Emden H, Rickelt S, Wirtz S, Scherer V (2008) A study on the validity of the multi-sphere discrete element method. Powder Technol 188:153–165. https://doi.org/10.1016/j.powtec.2008.04.037
115. Saeed MK, Siraj MS (2019) Mixing study of non-spherical particles using DEM. Powder Technol 344:617–627. https://doi.org/10.1016/j.powtec.2018.12.057
116. Markauskas D, Kačianauskas R (2011) Investigation of rice grain flow by multi-sphere particle model with rolling resistance. Granul Matter 13:143–148. https://doi.org/10.1007/s10035-010-0196-5
117. Wiacek J, Molenda M, Horabik J, Ooi JY (2012) Influence of grain shape and intergranular friction on material behavior in uniaxial compression: experimental and DEM modeling. Powder Technol 217:435–442. https://doi.org/10.1016/j.powtec.2011.10.060
118. Markauskas D, Kačianauskas R, Džiugys A, Navakas R (2010) Investigation of adequacy of multi-sphere approximation of elliptical particles for DEM simulations. Granul Matter 12:107–123. https://doi.org/10.1007/s10035-009-0158-y
119. Ma H, Zhao Y (2017) Modeling of the flow of ellipsoidal particles in a horizontal rotating drum based on DEM simulation. Chem Eng Sci 172:636–651. https://doi.org/10.1016/j.ces.2017.07.017
120. Zhou Z, Zou R, Pinson D, Yu A (2011) Dynamic simulation of the packing of ellipsoidal particles. Ind Eng Chem Res 50:9787–9798. https://doi.org/10.1021/ie200862n
121. Luo X, Zhao L, Zhang S et al (2018) Experimental and DEM studies on the distribution of voidage in the random filling bed of ellipsoidal particles. Powder Technol 340:400–410. https://doi.org/10.1016/j.powtec.2018.09.042
122. Ma H, Zhao Y, Cheng Y (2019) CFD-DEM modeling of rod-like particles in a fl uidized bed with complex geometry. Powder Technol 344:673–683. https://doi.org/10.1016/j.powtec.2018.12.066
123. Chen H, Zhao S, Zhou X (2019) DEM investigation of angle of repose for super-ellipsoidal particles. Particuology. https://doi.org/10.1016/j.partic.2019.05.005
124. Zhao Y, Xu L, Umbanhowar PB, Lueptow RM (2019) Discrete element simulation of cylindrical particles using super-ellipsoids. Particuology 46:55–66. https://doi.org/10.1016/j.partic.2018.04.007
125. Baran O, Eppinger T, Han K (2018) DEM simulation of cylinders and capsules in a fluidized bed. In: 8th world congress on particle technology, Orlando, pp 232–242
126. Zhong WQ, Zhang Y, Jin B, Zhang M (2009) Discrete element method simulation of cylinder-shaped particle flow in a gas-solid fluidized bed. Chem Eng Technol 32:386–391. https://doi.org/10.1002/ceat.200800516

127. Wu M, Peters B, Rosemann T, Kruggel-Emden H (2019) A forcing fictitious domain method to simulate fluid-particle interaction of particles with super-quadric shape. Powder Technol. https://doi.org/10.1016/j.powtec.2019.09.088
128. Lu G, Third JR (2012) Critical assessment of two approaches for evaluating contacts between super-quadric shaped particles in DEM simulations. Chem Eng Sci 78:226–235. https://doi.org/10.1016/j.ces.2012.05.041
129. Wang S, Fan Y, Ji S (2018) Interaction between super-quadric particles and triangular elements and its application to hopper discharge. Powder Technol 339:534–549. https://doi.org/10.1016/j.powtec.2018.08.026
130. Metzger MJ, Glasser BJ (2013) Simulation of the breakage of bonded agglomerates in a ball mill. Powder Technol 237:286–302. https://doi.org/10.1016/j.powtec.2012.12.006
131. Potyondy DO, Cundall PA (2004) A bonded-particle model for rock. Int J Rock Mech Min Sci 41:1329–1364. https://doi.org/10.1016/j.ijrmms.2004.09.011
132. Spettl A, Bachstein S, Dosta M et al (2016) Bonded-particle extraction and stochastic modeling of internal agglomerate structures. Adv Powder Technol 27:1761–1774. https://doi.org/10.1016/j.apt.2016.06.007
133. Park B, Min K, Thompson N, Horsrud P (2018) Three-dimensional bonded-particle discrete element modeling of mechanical behavior of transversely isotropic rock. Int J Rock Mech Min Sci 110:120–132. https://doi.org/10.1016/j.ijrmms.2018.07.018
134. Govender N, Wilke DN, Pizette P, Abriak N (2018) A study of shape non-uniformity and poly-dispersity in hopper discharge of spherical and polyhedral particle systems using the blaze-DEM GPU code. Appl Math Comput 319:318–336. https://doi.org/10.1016/j.amc.2017.03.037
135. Nezami EG, Hashash YMA, Zhao D, Ghaboussi J (2007) Simulation of front end loader bucket–soil interaction using discrete element method. Int J Numer Anal Methods Geomech 31:1147–1162. https://doi.org/10.1002/nag.594
136. Li X, Wang F, Zhang D et al (2019) Fluid-solid interaction simulation for particles and walls of arbitrary polygonal shapes with a coupled LBM-IMB-DEM method. Powder Technol 356:177–192. https://doi.org/10.1016/j.powtec.2019.08.006
137. Peña AA, Lind PG, Herrmann HJ (2008) Modeling slow deformation of polygonal particles using DEM. Particuology 6:506–514. https://doi.org/10.1016/j.partic.2008.07.009
138. Fu X, Yao Z, Zhang X (2017) Numerical simulation of polygonal particles moving in incompressible viscous fluids. Particuology 31:140–151. https://doi.org/10.1016/j.partic.2016.05.016
139. Džiugys A, Peters B (2001) An approach to simulate the motion of spherical and non-spherical fuel particles in combustion chambers. Granul Matter 3:231–265. https://doi.org/10.1007/PL00010918
140. Li J, Webb C, Pandiella SS, Campbell GM (2002) A numerical simulation of separation of crop seeds by screening – effect of particle bed depth. Food Bioprod Process Trans Inst Chem Eng Part C 80:109–117. https://doi.org/10.1205/09603080252938744
141. Locurto GJ, Zhang X, Zakirov V et al (1997) Soybean impacts: experiments and dynamic simulations. Trans ASAE 40:789–794. https://doi.org/10.13031/2013.21311
142. Zhang X, Vu-Quoc L (2000) Simulation of chute flow of soybeans using an improved tangential force-displacement model. Mech Mater 32:115–129. https://doi.org/10.1016/S0167-6636(99)00043-5
143. Xu T, Yu J, Yu Y, Wang Y (2018) A modelling and verification approach for soybean seed particles using the discrete element method. Adv Powder Technol 29:3274–3290. https://doi.org/10.1016/j.apt.2018.09.006
144. Sakaguchi E, Suzuki M, Favier JF, Kawakami S (2001) Numerical simulation of the shaking separation of Paddy and Brown rice using the discrete element method. J Agric Eng Res 79:307–315. https://doi.org/10.1006/jaer.2001.0706
145. Jiang G, Qiu B (2011) Discrete element method simulation of impact-based measurement of grain mass flow. In: 2011 international conference on computer distributed control and intelligent environmental monitoring. IEEE, Changsha, pp 419–422

146. Mousaviraad M, Tekeste MZ, Rosentrater KA (2017) Calibration and validation of a discrete element model of corn using grain flow simulation in a commercial screw grain auger. Trans ASABE 60:1403–1415. https://doi.org/10.13031/trans.12200
147. Coetzee CJ, Els DNJ (2009) Calibration of discrete element parameters and the modelling of silo discharge and bucket filling. Comput Electron Agric 65:198–212. https://doi.org/10.1016/j.compag.2008.10.002
148. Chung YC, Ooi JY (2008) A study of influence of gravity on bulk behaviour of particulate solid. Particuology 6:467–474. https://doi.org/10.1016/j.partic.2008.07.017
149. González-Montellano C, Ramírez A, Fuentes JM, Ayuga F (2012) Numerical effects derived from en masse filling of agricultural silos in DEM simulations. Comput Electron Agric 81:113–123. https://doi.org/10.1016/j.compag.2011.11.013
150. Markauskas D, Ramírez-Gómez Á, Kačianauskas R, Zdancevičius E (2015) Maize grain shape approaches for DEM modelling. Comput Electron Agric 118:247–258. https://doi.org/10.1016/j.compag.2015.09.004
151. Parafiniuk P, Molenda M, Horabik J (2013) Discharge of rapeseeds from a model silo: physical testing and discrete element method simulations. Comput Electron Agric 97:40–46. https://doi.org/10.1016/j.compag.2013.06.008
152. Raji AO, Favier JF (2004) Model for the deformation in agricultural and food particulate materials under bulk compressive loading using discrete element method. II: compression of oilseeds. J Food Eng 64:373–380. https://doi.org/10.1016/j.jfoodeng.2003.11.005
153. Wiacek J, Molenda M (2011) Moisture-dependent physical properties of rapeseed – experimental and DEM modeling. Int Agrophys 25:59–65
154. Wojtkowski M, Pecen J, Horabik J, Molenda M (2010) Rapeseed impact against a flat surface: physical testing and DEM simulation with two contact models. Powder Technol 198:61–68. https://doi.org/10.1016/j.powtec.2009.10.015
155. Molenda M, Horabik J, Łukaszuk J, Wiacek J (2011) Variability of intergranular friction and its role in DEM simulation of direct shear of an assembly of rapeseeds. Int Agrophys 25:361–368
156. Sarnavi HJ, Noor Mohammadi A, Modares Motlagh A, Rahmani Didar A (2013) Dem model of wheat grains in storage considering the effect of moisture content in direct shear test. Res J Appl Sci Eng Technol 5:829–841. https://doi.org/10.19026/rjaset.5.5029
157. Iroba KL, Mellmann J, Weigler F et al (2011) Particle velocity profiles and residence time distribution in mixed-flow grain dryers. Granul Matter 13:159–168. https://doi.org/10.1007/s10035-010-0222-7
158. Mellmann J, Iroba KL, Metzger T et al (2011) Moisture content and residence time distributions in mixed-flow grain dryers. Biosyst Eng 109:297–307. https://doi.org/10.1016/j.biosystemseng.2011.04.010
159. Patwa A, Ambrose RPK, Casada ME (2016) Discrete element method as an approach to model the wheat milling process. Powder Technol 302:350–356. https://doi.org/10.1016/j.powtec.2016.08.052
160. Sakai M, Yamada Y, Shigeto Y, Shibata K (2010) Large-scale discrete element modeling in a fluidized bed. Int J Numer Methods Fluids 64:1391–1335. https://doi.org/10.1002/fld.2364
161. Sakai M, Abe M, Shigeto Y et al (2014) Verification and validation of a coarse grain model of the DEM in a bubbling fluidized bed. Chem Eng J 244:33–43. https://doi.org/10.1016/j.cej.2014.01.029
162. Sakai M, Shigeto Y, Basinskas G et al (2015) Discrete element simulation for the evaluation of solid mixing in an industrial blender. Chem Eng J 279:821–839. https://doi.org/10.1016/j.cej.2015.04.130
163. Noid WG (2013) Perspective: coarse-grained models for biomolecular systems. J Chem Phys 139:090901. https://doi.org/10.1063/1.4818908
164. Klein ML, Shinoda W (2008) Large-scale molecular dynamics simulations of self-assembling systems. Science (80-) 321:798–800. https://doi.org/10.1126/science.1157834

165. Mori Y, Sakai M (2021) Visualization study on the coarse graining DEM for large-scale gas–solid flow systems. Particuology 59:24–33. https://doi.org/10.1016/j.partic.2020.07.001
166. Zhou L, Zhao Y (2021) CFD-DEM simulation of fluidized bed with an immersed tube using a coarse-grain model. Chem Eng Sci 231:116290. https://doi.org/10.1016/j.ces.2020.116290
167. Wang S, Shen Y (2022) Coarse-grained CFD-DEM modelling of dense gas-solid reacting flow. Int J Heat Mass Transf 184:122302. https://doi.org/10.1016/j.ijheatmasstransfer.2021.122302
168. Jia Y, Zhang Y, Xu J et al (2021) Coarse-grained CFD-DEM simulation to determine the multiscale characteristics of the air dense medium fluidized bed. Powder Technol 389:270–277. https://doi.org/10.1016/j.powtec.2021.05.014
169. Lungu M, Siame J, Mukosha L (2022) Coarse-grained CFD-DEM simulations of fluidization with large particles. Powder Technol 402:117344. https://doi.org/10.1016/j.powtec.2022.117344
170. Cundall PA (1988) Formulation of a three-dimensional distinct element model—Part I. A scheme to detect and represent contacts in a system composed of many polyhedral blocks. Int J Rock Mech Min Sci Geomech Abstr 25:107–116. https://doi.org/10.1016/0148-9062(88)92293-0
171. ESSS-Rocky (2018) Rocky-Dem Technical Manual. In: Models used in Rocky. ESSS, Rio de Janeiro, Brazil
172. Mack S, Langston P, Webb C, York T (2011) Experimental validation of polyhedral discrete element model. Powder Technol 214:431–442. https://doi.org/10.1016/j.powtec.2011.08.043
173. Höhner D, Wirtz S, Scherer V (2013) Experimental and numerical investigation on the influence of particle shape and shape approximation on hopper discharge using the discrete element method. Powder Technol 235:614–627. https://doi.org/10.1016/j.powtec.2012.11.004
174. Höhner D, Wirtz S, Scherer V (2015) A study on the influence of particle shape on the mechanical interactions of granular media in a hopper using the discrete element method. Powder Technol 278:286–305. https://doi.org/10.1016/j.powtec.2015.02.046
175. Oschmann T, Vollmari K, Kruggel-Emden H, Wirtz S (2015) Numerical investigation of the mixing of non-spherical particles in fluidized beds and during pneumatic conveying. Procedia Eng 102:976–985. https://doi.org/10.1016/j.proeng.2015.01.220
176. Potapov AV, Campbell CS (1996) A three-dimensional simulation of brittle solid fracture. Int J Mod Phys C 7:717–729. https://doi.org/10.1142/S0129183196000594
177. He M, Ren F, Cheng C (2018) Mechanism of strain burst by laboratory and numerical analysis. Shock Vib 2018:15. https://doi.org/10.1155/2018/8940798
178. Holt RM (2016) Bounds of elastic parameters characterizing transversely isotropic media: application to shales. Geophysics 81:c243–c252. https://doi.org/10.1190/geo2015-0485.1
179. Guo Y, Wassgren C, Curtis JS, Xu D (2018) A bonded sphero-cylinder model for the discrete element simulation of elasto-plastic fibers. Chem Eng Sci 175:118–129. https://doi.org/10.1016/j.ces.2017.09.029

Chapter 4
Computational Fluid Dynamic (CFD)

Abstract Theoretical fluid dynamics, experimental fluid dynamics, and computational fluid dynamic (CFD) are the three approaches for the study of problems in fluid dynamics. CFD as a flow simulation tool requires physical models described by the governing partial differential equations of fluid dynamics, appropriate accurate and stable numerical algorithms for the discretization of the governing equations on a mesh generated inside a chosen computational domain, and computer hardware for the solution of the discretized algebraic equations in the computational domain with appropriate boundary conditions. The discretized algebraic equations are coded to generate a computer code in a computer language (such as Fortran, Fortran 90, C, C++, Java, and Python). The computer code is debugged and validated by conducting benchmark simulations; this computer code is run on computer hardware to obtain the numerical solution of the flow field. In this context, CFD can be considered a multi-disciplinary field involving fluid dynamics, numerical mathematics, and computer science. CFD builds a bridge between the theoretical models and experiments since it is difficult to obtain the analytical solutions of the complex 3D fluid motions using the theoretical models, and it is costly to build the experimental apparatus and conduct experiments for a wide variety of flows of interest. As a result, CFD has become one of the most important technologies for conducting fluid dynamics research. Numerical simulations are also sometimes called "numerical experiments." However, unlike experiments, CFD can often provide detailed three-dimensional flow field data to analyze, design, and optimize industrial products such as the aerodynamic systems described in this book.

Keywords CFD · Discretization · Computational domain · Numerical algorithms · Turbulence modeling

4.1 Introduction

Computational fluid dynamic (CFD) solves the governing equations of fluid flow in an integral or differential form with appropriate boundary conditions in a suitably chosen computational domain to compute and visualize the desired flow field data

L. Zhou et al., *Discrete Element Method for Multiphase Flows with Biogenic Particles*, https://doi.org/10.1007/978-3-031-67729-8_4

for analysis, optimization, and improvements in product design. In the past three decades, many commercial and non-commercial codes have been developed for industrial applications. Among the commercial codes, the most well-known are Ansys Fluent, CFX, COMSOL, etc. Among the non-commercial open-source codes are OpenFOAM, SU2, and others. There are also many industries and government-funded proprietary codes in many countries worldwide. A CFD code consists of the following four building blocks: (a) geometry modeling tool, (b) mesh generation software, (c) flow solver, and (d) numerical data processing and flow visualization tool. For numerical simulations of the flow field of a typical aerodynamic system described in Chap. 3 and cyclones described in Chap. 7, a brief description of the building blocks of CFD is given in the following sections.

4.2 Geometry Modeling

The three-dimensional geometry of the aerodynamic systems described in Chap. 3 and the cyclone described in Chap. 7 is modeled by UG software such as Nx, AutoCAD, and SolidWorks. Then, the simulation domains for computations are extracted from it. Finally, all the solid domains of the aerodynamic system for simulation are assembled. The geometry modeling process is relatively straightforward.

4.3 Mesh Generation

The entire computational domain for aerodynamic systems described in Chap. 3 and cyclone described in Chap. 7 is meshed using a mesh generation software, for example, Ansys, which contains Gambit, ICEM, Design Modeler, and Space Claim for geometry modeling and mesh generation, and can automatically generate a 3D single or multi-block structured, unstructured, or hybrid grid with a desired topology for very complex external and internal flow configurations. Since mesh generation is important for accurate simulation, the mesh type, topology, and density in various regions of the computational domain should be carefully chosen for accurate and computationally efficient calculations. The refined tetrahedral meshes are employed to simulate the aerodynamic systems described in Chap. 3 and the cyclones described in Chap. 7. The mesh is refined in the boundary layer regions close to the solid surfaces. The automatic near-wall treatment is employed to suit a turbulence model's requirements, depending on whether the turbulence model is a low Reynolds number (LRN) model that can integrate into the wall or is a high Reynolds number (HRN) model that requires wall functions. The use of wall functions does not require very fine mesh with y^+ of O (1) near the wall; usually y^+ of (100) when employing the wall functions requiring less dense mesh near the wall.

4.4 Governing Equations

CFD has become an essential tool for simulating the flow field in fluid systems, both internal and external [1]. Its expanding scope of applications has opened new vistas for solving engineering problems in many industries, including agriculture. Employing mathematical modeling based on partial differential equations, numerical methods such as the finite volume method for solving the partial differential equations, and a variety of other software tools for geometry modeling and mesh generation and post-processing of data, CFD can provide quantitative predictions of fluid flows by solving the transport equations of mass, momentum, and energy [2]. Industries have widely used this technology to optimize design processes, minimize energy costs, and develop new designs without extensive experimentation. CFD has proved helpful in predicting fluid flow patterns in processes such as mixing, drying, separation, and fluidized bed granulators [3]. Typically, to study the continuum flow pattern in the presence of a secondary particulate phase, the system domain is divided into small grid cells, and the volume-averaged Navier-Stokes equations govern the motion of the fluid phase [4–7]. The continuity and momentum conservation equations can be expressed using Eqs. (4.1) and (4.2), respectively. The finite volume method has emerged as the preferred method for solving these equations.

$$\frac{\partial\left(\rho_f \alpha_f\right)}{\partial_t} + \nabla.\left(\rho_f \alpha_f V_f\right) = 0 \tag{4.1}$$

$$\frac{\partial\left(\rho_f \alpha_f v_f\right)}{\partial_T} + \nabla.\left(\rho_f \alpha_f V_f V_f\right) = -\propto_f \nabla p + \nabla.\left(\propto_f \tau_f\right) + \propto_f \rho_f g - F_i^f \tag{4.2}$$

where α_f is the fluid volume fraction in each cell, ρ_f is the fluid density, v_f is the average velocity of a fluid cell, p is the pressure shared by the two phases, τ_f is the viscous stress tensor, and F_i^f is the interaction force of the fluid on the particle i.

Efficient and reliable numerical models and settings are essential to obtain accurate results in CFD simulations of aerodynamic systems. These include selecting an appropriate grid size, turbulence model, and boundary conditions for the application. To achieve this, multiple numerical calculations with varying mesh sizes and turbulence models need to be conducted to determine the best models and numerical settings for agricultural aerodynamic systems [1]. The computed solutions should be grid-independent; therefore, suitable grids are essential to obtain accurate and computationally efficient solutions [3]. The choice of turbulence model for internal flow in agricultural aerodynamic systems depends on several factors, including the physics involved, the accuracy required, and the computational resources available, including the simulation cost [8]. As such, it is not feasible to adopt a single turbulence model for all problems, and a thorough analysis of the flow geometry and parameters should be conducted to determine the best turbulence model for each application. Fortunately, recent advances in CFD techniques, turbulence modeling, and computational processing equipment have enabled the modeling of complex turbulent fluid systems. In addition, commercial CFD software such as STAR-CCM+ from CD-Adapco, COMSOL Multiphysics, SIMULIA PowerFlow, ANSYS Fluent, and other open or licensed software sources offer parallel computing options to speed up the computational processes. These software codes support a wide range of turbulence models, ranging from the k-ϵ, renormalization group (RNG)-based k-ϵ, k-ω, and SST-based k-ω models to more complex models such as the full Reynolds stress model (RSM), providing choices for a closer and more realistic representation of the velocity, pressure, and temperature fields of the fluid in the given domain [2]. While it is impossible to determine the best turbulent model for a specific application, this chapter, for the sake of completeness, briefly reviews the most popular turbulence models used in aerodynamic systems related to agricultural applications. By carefully selecting the numerical settings and turbulence model, it is possible to obtain accurate and reliable CFD simulations for optimizing the system's performance.

4.5 Numerical Algorithms

Before conducting turbulent flow simulations in a specific aerodynamic system, the computational domain of the system needs to be appropriately discretized into many discrete grid elements. Then, the governing Reynolds-averaged Navier-Stokes (RANS) equations derived from Eqs. (4.1) and (4.2) in conjunction with the turbulence modeling partial differential equations (PDEs) need to be discretized on these grid elements following an appropriately stable and convergent numerical algorithm; as a result, all the PDEs transform into non-linear algebraic equations that can be linearized using an iterative scheme. The linearized algebraic equations are then solved by an efficient matrix algebra algorithm, depending on the arrangement of the matrix elements.

At present, there are three commonly used discretization methods, which include the finite difference method, the finite element method, and the finite volume method [9]. The finite difference method is an early discretization method that was successfully used primarily for flow problems with simple geometries. Its basic feature is to replace the derivative terms in the governing equations with the corresponding difference expressions at each grid node created by the intersection of the three grid lines representing the three coordinate directions. The finite-difference equations result in algebraic equations at each node, and the algebraic equations at each node include the unknown flow variables at this node and from neighboring nodes. The required numerical solutions for the flow variables are then obtained by solving these algebraic equations with appropriate boundary conditions. The main idea behind the finite element method is to divide the computational domain into a series of small elements, add several points to each element as nodes, and then obtain the discrete equations by integrating the governing equations on these finite elements.

The basic idea of the finite control volume method is to divide the computational domain into many control volumes, and a node represents each control volume. The discrete equations are derived by integrating the governing equations over the control volume. In this process, some assumptions are made about the function representing the control volume—linear, quadratic, or cubic and its first-order derivatives, which are required at the interface of the two control volumes. Thus, different formulations for various numerical algorithms have been generated in the past 50 years. The finite volume method is the most widely used method in commercial CFD codes (ANSYS, COMSOL, CFX, etc.), non-commercial open-source CFD codes (OpenFOAM, SU2, etc.), and proprietary industrial and government codes. In this book, the focus is on the commercial code ANSYS Fluent/CFX.

In incompressible flow, there is no pressure term in the continuity equation, and therefore, pressure and velocity are not strongly coupled in the governing equations. This problem is addressed using the so-called "coupled" or "uncoupled" algorithm. Here, the "coupled" algorithm refers to the pressure-velocity coupling algorithm of Launder and Spalding [10] (also known as the pressure correction approach), and the non-coupled algorithm refers to the solution method requiring the solution of the Poisson equation for pressure along with the momentum equations. The "Pressure Correction"-based algorithms are known as the SIMPLE (Semi-Implicit Method for Pressure Linked Equations) family of algorithms in the literature [11]. SIMPLEC is highly recommended among the SIMPLE family of algorithms since the influence of the velocity correction values of the adjacent grid points is omitted, making the algorithm simpler; it also has good convergence properties [12]. Other variations of the algorithm are not discussed here, and the reader is referred to Refs. [13, 14].

The high-order (third-order) upwind scheme is generally used for the convection terms, and the central difference scheme is used for the diffusion terms in the RANS and turbulence modeling equations. The unsteady second-order implicit time integration scheme is used for the time-dependent term in the transient flow simulations.

After the discrete algebraic equations are established by the finite volume method using the "coupled" or "uncoupled" solution approach, an "implicit," "explicit-implicit," or "explicit" solution technique can be used to solve the equations. In the implicit method, all equations are solved simultaneously for the flow variables u, v, w, and p. The solution can also be obtained using an explicit–implicit method in which the solution algorithm has both the explicit and implicit parts; the values of the flow variables in the explicit parts are known, while all flow variables are computed simultaneously in the implicit part. In the explicit solution algorithm, the algebraic equations are solved sequentially, one by one at a time. A particular algorithm is chosen based on the memory requirements and its accuracy and stability on a given mesh defined in terms of the CFL (Courant–Fredrick–Levy) condition or Courant number. The CFL condition for the implicit or explicit-implicit algorithm may be larger than that for the purely explicit algorithm; however, the implicit algorithm is more complex since it requires matrix inversion and substantially more memory requirements than the explicit algorithm.

4.6 Boundary Conditions

The boundary conditions are determined based on the aerodynamic system's operating conditions. The velocity inlet boundary condition is used at the airflow inlet; the velocity distribution at the inlet is assumed to be uniform and is determined by the flow rate. The outlet boundary condition is set as a pressure outlet. The no-slip boundary condition is employed on all solid walls of the system. The enhanced wall function is used near the wall region to compute turbulent flow efficiently [15]. The solution is considered converged for steady-state computations if the root-mean-square residuals of all governing equations and flow variables drop below 10^{-5}. In the case of transient simulations, for example, for a cyclone, when the root-mean-square residuals drop below 10^{-5}, the convergence of each physical time step is obtained in 4–10 iterations.

4.7 Turbulence Models

4.7.1 *Wray-Agarwal One-Equation Model*

The Wray-Agarwal (WA) turbulence model is a one-equation linear eddy-viscosity model derived from the *k-ω* closure [16]. It has been shown to have improved accuracy over the one-equation Spalart-Allmaras (SA) model [17] and is competitive with the two-equation SST *k-ω* model [18] in many 2D and 3D complex flows. It behaves like the two-equation *k-ω* model in the near wall region and the standard two-equation *k-ε* model in the free stream and shear flow regions. The inclusion of the cross-diffusion term in the R ($=k/\omega$) equation and a blending function that allows smooth switching between the two destruction terms are the most innovative features of this model. The transport equation for the variable R in the WA model is as follows:

$$\frac{\partial R}{\partial t}+\frac{\partial u_j R}{\partial x_j}=\frac{\partial}{\partial x_j}\left[(\nu+\sigma_R R)\frac{\partial R}{\partial x_j}\right]+C_1 RS+f_1 C_{2k\omega}\frac{R}{S}\frac{\partial R}{\partial x_j}\frac{\partial S}{\partial x_j} -(1-f_1)\min\left[C_{2k\varepsilon}\left(\frac{\frac{\partial S}{\partial x_j}\frac{\partial S}{\partial x_j}}{S^2}\right),C_m\frac{\partial R}{\partial x_j}\frac{\partial R}{\partial x_j}\right] \tag{4.3}$$

where S is the mean strain given by:

$$S=\sqrt{2S_{ij}S_{ij}}\ \text{where}\ S_{ij}=\frac{1}{2}\left(\frac{\partial u_i}{\partial x_j}+\frac{\partial u_j}{\partial x_i}\right) \tag{4.4}$$

The turbulent kinematic eddy viscosity is given by the equation:

$$\nu_T = f_\mu R \tag{4.5}$$

The wall-blocking effect is accounted for by the damping function f_μ given by:

$$f_\mu = \frac{\chi^3}{\chi^3 + C_w^3}, \chi = \frac{R}{\nu} \tag{4.6}$$

The switching function f_1 allows the WA model to behave as a one-equation k-ω or one-equation k-ε model and is defined as:

$$f_1 = \min\left(\tan h\left(\arg_1^4\right), 0.9\right) \tag{4.7}$$

$$\arg_1 = \frac{1 + \frac{d\sqrt{RS}}{\nu}}{1 + \left[\frac{\max\left(d\sqrt{RS}, 1.5R\right)}{20\nu}\right]^2} \tag{4.8}$$

where d is the minimum distance to the nearest wall. The various constants in the WA model are given as; $C_1 = f_1(C_{1k\omega} - C_{1k\varepsilon}) + C_{1k\varepsilon}$, $C_{1k\omega} = 0.0829$, $C_{1k\varepsilon} = 0.1127$, $C_\omega = 8.54$, $\sigma_R = f_1(\sigma_{k\omega} - \sigma_{k\varepsilon}) + \sigma_{k\varepsilon}$, $\sigma_{k\omega} = 0.72$, $\sigma_{k\varepsilon} = 1.0$, $\kappa = 0.41$, $C_m = 8.0$

$$C_{2k\omega} = \frac{C_{1k\omega}}{\kappa^2} + \sigma_{kw}, C_{2k\varepsilon} = \frac{C_{1k\varepsilon}}{\kappa^2} + \sigma_{k\varepsilon}$$

This model has been extended to include the effects of rotation and curvature (R&C) and has demonstrated success in computing 2D and 3D flows with rotation and curvature [19], including the flow in a cyclone separator described in Chap. 7.

4.7.2 Standard k-ε Model

The k-ε model, initially proposed by Launder and Spalding [10, 20], is widely used in industrial flows and heat transfer simulations due to its combination of accuracy, validity, economy, and realism for a broad range of turbulent flows. The model is based on two transport equations for turbulence kinetic energy (k) and its dissipation rate (ε) as given by Eqs. (4.9) and (4.10), respectively. It assumes that the flow is fully turbulent with a HRN and a negligible effect of molecular viscosity [8, 21]. However, the standard k-ε model's accuracy is compromised in modeling high internal velocity with high swirl and turbulence (anisotropic flow), rendering it inadequate for modeling real flows in aerodynamic systems, especially the cyclone separators [2, 22–24]. As a result, alternative models are needed that can simulate complex flow phenomena more accurately, for example, in cyclone separators and other turbulent flow systems.

$$\frac{\partial}{\partial t}\left(\rho_f k\right)+\frac{\partial}{\partial x_j}\left(\rho_f k u_{if}\right)=\frac{\partial}{\partial x_j}\left[\left(\mu_f+\frac{\mu_t}{\sigma_k}\right)\frac{\partial k}{\partial x_j}\right]+G_k+G_b-\rho\varepsilon-Y_M+S_k \tag{4.9}$$

$$\frac{\partial}{\partial t}\left(\rho_f \varepsilon\right)+\frac{\partial}{\partial x_j}\left(\rho_f \varepsilon u_{if}\right)=\frac{\partial}{\partial x_j}\left[\left(\mu_f+\frac{\mu_t}{\sigma_\varepsilon}\right)\frac{\partial \varepsilon}{\partial x_j}\right]+C_1\frac{\varepsilon}{k}\left(G_k+C_3 G_b\right)-C_2\rho_f\frac{\varepsilon^2}{k}+S_\varepsilon \tag{4.10}$$

where u_f is the average velocity component of the fluid at position x, μ_f is the turbulent viscosity, G_k represents the generation of turbulent kinetic energy due to the average velocity gradients, G_b represents the generation of turbulent kinetic energy due to buoyancy, Y_M represents the contribution of the fluctuating dilatational incompressible turbulence to the overall dissipation rate, and σ_k and σ_ε are the turbulent Prandtl numbers corresponding to k and ε, respectively. S_k and S_ε are user-defined source terms; R_ε represents the dissipation of swirl and rotational effect on turbulence; and C_1, C_2, and C_3 are constants.

4.7.3 *RNG* k-ε *Model*

The standard *k-ε* model has long been plagued by several limitations; however, a solution was proposed in the form of the RNG-based *k-ε* model. This model is based on the Navier-Stokes equations and employs the RNG technique to overcome the shortcomings of its predecessor [25]. The most remarkable feature of the RNG-based *k-ε* model is its ability to consider the effect of swirl on turbulence, which has eluded previous models. It also includes an additional dissipation rate equation term that vastly improves the accuracy of rapidly strained flows. The model's effective viscosity (μ_{eff}) is derived analytically, providing a differential formula for LRN effects. In addition, the RNG theory provides an analytical formula for turbulent Prandtl numbers, while the standard *k-ε* model uses user-specified, constant values. While the standard *k-ε* model is a HRN model, the RNG theory provides an analytically derived differential formula for effective viscosity that accounts for LRN effects. However, effective use of this feature does depend on an appropriate treatment of the near-wall region. Due to these refinements, the RNG-based *k-ε* model is more accurate and reliable than the standard *k-ε* model and is suitable for a wider range of flow types. Several studies have shown that the model outperforms its predecessor in accuracy and predictive capabilities [8, 26, 27]. In short, the RNG-based *k-ε* model has been a major step forward in turbulence modeling; its governing transport equations for k and ε are given by Eqs. (4.11) and (4.12), respectively.

$$\frac{\partial}{\partial t}(\rho_f k)+\frac{\partial}{\partial x_j}(\rho_f k u_{if})=\frac{\partial}{\partial x_j}\left[\alpha_k \mu_{f_{\text{eff}}}\frac{\partial k}{\partial x_j}\right]+G_k+G_b-\rho\varepsilon-Y_M+S_k \tag{4.11}$$

$$\frac{\partial}{\partial t}(\rho_f \varepsilon)+\frac{\partial}{\partial x_j}(\rho_f \varepsilon u_{if})=\frac{\partial}{\partial x_j}\left[\alpha_\varepsilon \mu_{\text{eff}}\frac{\partial \varepsilon}{\partial x_j}\right]+C_1\frac{\varepsilon}{k}(G_k+C_3G_b)-C_2\rho_f\frac{\varepsilon^2}{k}-R_\varepsilon+S_\varepsilon \tag{4.12}$$

In Eqs. (4.11) and (4.12), $\mu_{f_{\text{eff}}}$ is the effective turbulent viscosity and α_k and α_ε are the inverse effective Prandtl numbers corresponding to k and ε, respectively. For more details and the background theory behind the RNG k-ε model, the reader refers to Refs. [28–30].

4.7.4 SST k-ω Model

Menter's shear stress transport (SST) turbulence model [18] is a widely used, robust two-equation eddy-viscosity turbulence model. The model combines the features of the k-ω turbulence model and the k-ε turbulence model such that the k-ω model is used in the inner region of the boundary layer and switches to the k-ε model in the free shear flow region. The formulation of the SST model is based on information from many experiments and attempts to predict solutions to typical engineering problems. The following equations provide details about the model:

$$\frac{\partial(\rho_f k)}{\partial t}+\frac{\partial(\rho_f u_{if} k)}{\partial x_j}=P-\beta^*\rho_f\omega k+\frac{\partial}{\partial x_j}\left[(\mu+\sigma_k\mu_t)\frac{\partial k}{\partial x_j}\right] \tag{4.13}$$

$$\frac{\partial(\rho_f\omega)}{\partial t}+\frac{\partial(\rho_f u_{if}\omega)}{\partial x_j}=\frac{\gamma}{v_t}P-\beta\rho_f\omega^2+\frac{\partial}{\partial x_j}\left[(\mu+\sigma_\omega\mu_t)\frac{\partial\omega}{\partial x_j}\right]+2(1-F_1)\frac{\rho_f\sigma_{\omega 2}}{\omega}\frac{\partial k}{\partial x_j}\frac{\partial\omega}{\partial x_j} \tag{4.14}$$

$$P=\tau_{ij}\frac{\partial u_{if}}{\partial x_j} \tag{4.15}$$

The turbulence eddy viscosity is computed from

$$v_t=\frac{\rho a_1 k}{\max(a_1\omega,\Omega F_2)} \tag{4.16}$$

Each of the constants is a blend of an inner (subscript 1) and an outer (subscript 2) constant via

$$\phi = F_1\phi_1 + (1 - F_1)\phi_2 \tag{4.17}$$

Additional functions are defined as follows:

$$F_1 = \tan h\left(\arg_1^4\right) \tag{4.18}$$

$$\arg_1 = \min\left[\max\left(\frac{\sqrt{k}}{\beta^* \omega d}, \frac{500\nu}{d^2\omega}\right), \frac{4\rho_f \sigma_{\omega 2} k}{CD_{k\omega} d^2}\right] \tag{4.19}$$

$$CD_k = \max\left(2\rho_f \sigma_{\omega 2} \frac{1}{\omega}\frac{\partial k}{\partial x_j}\frac{\partial \omega}{\partial x_j}, 10^{-20}\right) \tag{4.20}$$

$$F_2 = \tan h\left(\arg_2^2\right) \tag{4.21}$$

$$\arg_2 = \max\left(2\frac{\sqrt{k}}{\beta^* \omega d}, \frac{500\nu}{d^2\omega}\right) \tag{4.22}$$

4.7.5 Reynolds Stress Model (RSM)

The full RSM is a powerful turbulence model developed to accurately capture the intricate effects of streamline curvature, swirl, rotation, and rapid changes in the strain rate, presumably overcoming the limitations of one-equation and two-equation turbulence models [31, 32]. This model holds great potential in providing better predictions for complex flows, especially those exhibiting heavy swirls and anisotropic turbulence within aerodynamic systems like cyclone devices [28, 33–36]. The algebraic transport equation of the Reynolds stress tensor, $\rho u_i u_j$, is elegantly expressed by Eq. (4.23), with a comprehensive description laid out in Eqs. (4.24), (4.25), (4.26), (4.27), (4.28), (4.29), (4.30), and (4.31) [37–39]. By utilizing RSM, one can better understand flow dynamics and optimize design and operational parameters to achieve the desired performance.

$$\frac{\partial}{\partial t}\left(\rho_f \overline{u_i' u_j'}\right) + C_{ij} = D_{T,ij} + D_{L,ij} + P_{ij} + G_{ij} + \phi_{ij} + \varepsilon_{ij} + F_{ij} + S_{\text{user}} \tag{4.23}$$

where the first term on the left side of Eq. (4.23) is the local time derivative of the Reynolds stress tensor; C_{ij} represents the convection term; $D_{T,ij}$ represents the turbulent diffusion term; $D_{L,ij}$ represents the molecular diffusion term; P_{ij} and G_{ij} represent the stress and buoyancy production rates of $u_i' u_j'$, respectively; ϕ_{ij} represents the pressure strain consisting of three contributions to the pressure-strain correlation; ε_{ij} is the dissipation; F_{ij} is the production by system rotation; and S_{user} is the user-defined source term. The various terms on the right-hand side of Eq. (4.24) are described as follows.

$$C_{ij} = \frac{\partial}{\partial x_k}\left(\rho_f u_k \overline{u_i' u_j'}\right) \tag{4.24}$$

$$D_{L,ij} = \frac{\partial}{\partial x_k}\left[\mu \frac{\partial}{\partial x_k}\left(\overline{u_i' u_j'}\right)\right] \tag{4.25}$$

$$P_{ij} = -\rho_f\left(\overline{u_i' u_k'}\frac{\partial u_j}{\partial x_k} + \overline{u_j' u_k'}\frac{\partial u_i}{\partial x_k}\right) \tag{4.26}$$

$$F_{ij} = -2\rho_f \Omega_k\left(\overline{u_j' u_m'}\varepsilon_{ikm} + \overline{u_i' u_m'}\varepsilon_{jkm}\right) \tag{4.27}$$

$$D_{T,ij} = -\frac{\partial}{\partial x_k}\left(\frac{\mu_t}{\sigma_k}\frac{\partial \overline{u_i' u_j'}}{\partial x_k}\right) \tag{4.28}$$

$$P_{ij} = \rho_f\left(\overline{u_i' u_k'}\frac{\partial u_j}{\partial x_k} + \overline{u_j' u_k'}\frac{\partial u_i}{\partial x_k}\right) \tag{4.29}$$

$$\epsilon_{ij} = 2\mu_t \overline{\frac{\partial u_i'}{\partial x_k}\frac{\partial u_j'}{\partial x_k}} \tag{4.30}$$

$$\phi_{ij} = \phi_{ij,1} + \phi_{ij,2} + \phi_{ij,3} \tag{4.31}$$

where $\phi_{ij,1}$ in Eq. (4.20) is the slow pressure strain, which represents a return to isotropy by purely turbulent interactions [40]; $\phi_{ij,2}$ is the rapid pressure strain taking account of the return to isotropy by the interaction between the turbulence and the mean velocity gradients [32]; and $\phi_{ij,3}$ is the wall-reflection term, which is responsible for damping the velocity fluctuation component close and perpendicular to the wall, and is included by default in the RSM. Each component of the pressure strain can be formulated as described in Eqs. (4.32), (4.33), and (4.34):

$$\phi_{ij,1} = -C_1 \rho_f \frac{\epsilon}{k}\left[\overline{u_i' u_j'} - \frac{2}{3}\delta_{ij} k\right] \tag{4.32}$$

$$\phi_{ij,2} = -C_2\left[P_{ij} - \frac{2}{3}\delta_{ij}\left(-\overline{u_i' u_j'}\frac{\partial u_i}{\partial x_j}\right)\right] \tag{4.33}$$

$$\phi_{ij,3} = C_1' \frac{\epsilon}{k}\left(\overline{u_k' u_m'} n_k n_m \delta_{ij} - \frac{3}{2}\overline{u_i' u_k'} n_k n_j - \frac{3}{2}\overline{u_j' u_k'} n_k n_i\right)\frac{C_l k^{\frac{3}{2}}}{\varepsilon d}$$
$$+C_2'\left(\phi_{km,2} n_k n_m \delta_{ij} - \frac{3}{2}\phi_{ik,2} n_k n_j - \frac{3}{2}\phi_{jk,2} n_k n_i\right)\frac{C_l k^{3/2}}{\varepsilon d} \tag{4.34}$$

where n_k is the x_k component of the unit normal to the wall; d is the normal distance to the wall; and $C_l = \frac{C_\mu^{3/4}}{k}$. Due to system rotation, the following terms can be neglected in some applications: convection, molecular diffusion, and production. For isothermal flows, buoyancy production can also be neglected.

4.7.6 Large Eddy Simulation (LES)

A large eddy simulation (LES) turbulence model is used to capture the velocity field, which is a prominent variable in the cyclone separator case due to the anisotropic and chaotic turbulence. The governing Navier-Stokes equations describing the conservation of mass and momentum are solved in each computational cell for the gas phase. The grid cell scale directly resolves the largest eddies in LES simulations. For smaller eddies (sub-grid-scale eddies), a turbulence model (sub-grid-scale model) is used based on the universal behavior of turbulence in and beyond the inertial subrange, and since the grid determines which eddies are resolved and which have to be modeled [35]. Therefore, the LES approaches require fine grids and small-time steps, particularly for wall-bounded flows. However, it can give details on the structure of turbulent flows and is computationally cheaper than the RSM [41]. Besides, the RSM is poor at predicting velocity fluctuations and collection efficiency, especially for small particles, due to the vortex core phenomenon's processing [42]. The pressure and velocity values can be obtained by defining gas properties such as density and viscosity. Hence, the governing equations of the continuous gas phase are given in Eqs. (4.35) and (4.36), respectively [43, 44].

$$\frac{\partial\left(\rho_g \alpha_g\right)}{\partial_t} + \frac{\partial\left(\rho_g \alpha_g u_i\right)}{\partial_{x_i}} = 0 \tag{4.35}$$

$$\frac{\partial\left(\rho_g \alpha_g v_g\right)}{\partial_t} + \frac{\partial\left(\rho_g \alpha_g u_i u_i\right)}{\partial_{x_i}} = -\propto_g \nabla p + \alpha_g \frac{\partial}{\partial_{x_j}}\left(\tau_{\mu_{ij}} + \tau_{d_{ij}} + \tau_{t_{ij}}\right) + \propto_g \rho_g g - \frac{1}{\rho_g} F_i^g \tag{4.36}$$

where $\tau_{\mu_{ij}}$ is viscosity stress tensor, $\tau_{d_{ij}}$ is the drift tensor, and $\tau_{t_{ij}}$ is the eddy viscosity stresses tensor effect on the grid scale where these tensors can be calculated as given in Eqs. (4.37), (4.38), and (4.39) [42],

$$\tau_{\mu_{ij}} = 2\mu_g \overline{S_{ij}} \tag{4.37}$$

$$\tau_{d_{ij}} = \sum_{p=1}^{n} \rho_p \alpha_p u_{p_i} u_{p_j} \tag{4.38}$$

$$\tau_{t_{ij}} = \rho_g l_s^2 \sqrt{2\overline{S_{ij}}\,\overline{S_{ij}}} \left(\frac{\partial u_i}{\partial x_j} + \frac{\partial u_j}{\partial x_i} \right) \tag{4.39}$$

where $\overline{S_{ij}}$ is the mean strain rate tensor based on resolved scales and l_s is the intermediate local grid.

4.7.7 Detached Eddy Simulation (DES)

The detached eddy simulation (DES) is a modification of the original RANS turbulence model in which the model switches to a sub-grid-scale formulation in regions fine enough for LES calculations. The regions near the solid boundaries and where the turbulent length scale is less than the maximum grid dimension are assigned the RANS mode of solution, but as the turbulent length scale exceeds the grid dimension, the regions are solved using the LES mode. Therefore, the grid resolution is not as demanding as pure LES, thereby considerably cutting down on the cost of the computations with increased accuracy compared to RANS simulation. The DES model can be implemented with any RANS model described above by appropriately modifying the length scale explicitly or implicitly involved in the RANS turbulence model [45, 46]. Other modifications to DES, namely, the delayed detached eddy simulation and implicit delayed detached eddy simulations, have been proposed in the literature to improve the accuracy and efficiency of DES for some flows.

4.7.8 Near-Wall Treatment

The turbulence modeling near the wall substantially affects the accuracy of the numerical solutions since the walls are the primary source of vorticity and turbulence. The flow variables have significant gradients in the near-wall region, and the transport of momentum and other scalar quantities are most important. Consequently, a precise depiction of the flow in the near-wall region is crucial in ensuring successful predictions of wall-bounded turbulent flows. For fully developed turbulent flow, numerous experiments have shown that the near-wall region can be subdivided into three layers, namely, the viscous sublayer, the transition or buffer layer, and the logarithmic layer. The viscous sublayer is almost laminar, with the velocity component parallel to the wall having a linear distribution along the direction normal to the wall. The transition or buffer layer is located outside the viscous sublayer, where the influence of viscosity is comparable to the action of the turbulent shear stress. The flow conditions are more complex in this layer, and due to its small thickness, it is often not separately delineated in engineering calculations but is included in the logarithmic layer. The logarithmic layer is the outermost layer where turbulence is fully developed, the velocity profiles have logarithmic behavior, and the influence of molecular viscosity is insignificant; turbulent shear stress dominates the flow field. To describe the velocity profile in the viscous sublayer and the logarithmic layer, two dimensionless parameters y^+ and u^+ are introduced to represent the velocity and the wall distance, defined as:

$$u^+ = \frac{u}{u_\tau} \tag{4.40}$$

$$y^+ = y\frac{u_\tau}{\rho} \tag{4.41}$$

where ρ is density and u_τ is the shear velocity (or friction velocity), which can be written as:

$$u_\tau = \sqrt{\frac{\tau}{\rho}} \tag{4.42}$$

The shear stress τ is proportional to the velocity gradient,

$$\tau = \mu\frac{du}{dy} \tag{4.43}$$

where μ is the molecular viscosity of the fluid.

In the viscous sublayer ($0 < y^+ < 5$), near the wall, the viscous effects dominate the flow, and the velocity profile is linear, given by:

$$u^+ = y^+ \tag{4.44}$$

In the buffer layer ($5 < y^+ < 30$), which serves as a transition layer between the viscous sublayer and the log law layer, the viscous force due to molecular viscosity is close to the turbulent shear stress. It cannot be described by a simple formula. However, in most flow problems, its thickness is very small, it does not affect the flow field, and it is generally included in the log-law layer. The log-law layer ($30 < y^+ < 300$), where the velocity profile follows a logarithmic behavior up to the end of the boundary layer, can be expressed as:

$$u^+ = \frac{1}{\kappa} \log E y^+ + C^+ \tag{4.45}$$

where $\kappa \approx 0.41$ is the von Kármán constant, and $E \approx 9.8$; these are empirical values that can vary slightly from case to case. C^+ is also an empirical parameter, and it changes according to the wall roughness (e.g., $C^+ \approx 5$ for smooth walls). In the ANSYS software, the log law is generally employed when $y^+ > 11.225$, and the laminar stress–strain relationship is applied when $y^+ < 11.225$.

There are two main approaches to dealing with the near-wall region in the numerical simulation: the first approach is not to directly solve for the viscous sublayer and the buffer layer, but to solve from the viscous sublayer to the outer layer of the turbulent boundary layer by using a semi-empirical formula for the entire boundary layer and employing the so-called wall function method (this method is employed with HRN turbulence models); and the second approach is to directly solve to the wall including the laminar sublayer layer, buffer layer, and the log layer and analyze the influence of turbulent eddy viscosity to the wall (this method is employed with LRN turbulence models).

For many HRN flows, the wall function method can save a lot of computational resources. This is because the gradients in the near-wall region are large in the flow variables. Therefore, many more grid nodes are required for accurate resolution of the flow field, while only a few grid nodes are required when using the wall functions in the near-wall region. For many 3D industrial applications, such as flows in the hydraulic machinery, the wall function method is often employed to obtain numerical solutions of acceptable engineering accuracy for quantities of interest, for example, pump head and efficiency of a mixed flow pump, using substantially less computational resources. The commonly used wall functions include the so-called standard wall function and the enhanced wall function; the details are provided in the book by Wilcox [47].

When the LRN turbulence model is employed, the flow field in the viscous sublayer is solved directly, which requires not only the dimensionless distance $y+ < 1$ from the first layer of the grid to the wall but also at least 15–20 layers of the grid in the near wall region to accurately calculate the velocity and skin friction near the wall.

4.8 Summary

The CFD has become an indispensable tool for engineers and researchers alike. It aids in understanding, analyzing, and optimizing the complex flow patterns found in various aerodynamic systems and specialized equipment such as cyclones, as outlined in Chap. 7. Over the past three decades, significant commercial and open-source CFD software advancements have enabled the detailed simulation of fluid flow phenomena. This has dramatically benefited industries, allowing them to optimize processes, minimize energy consumption, and enhance product designs without resorting to costly physical experiments. Conducting CFD simulations involves four essential building blocks: geometry modeling, mesh generation, defining governing equations, and selecting numerical algorithms. Tools such as Nx, AutoCAD, and SolidWorks are used for geometry modeling, enabling the creation of intricate 3D models of aerodynamic systems that are then meshed using specialized software to define the computational domain. The governing equations, typically the Navier-Stokes equations, are discretized using finite volume methods. This process converts partial differential equations into algebraic equations suitable for computation. Central to achieving accurate simulations is the selection of turbulence models, each offering varying levels of accuracy and computational efficiency. Among the commonly used turbulence models are the Wray-Agarwal one-equation Model with rotation-curvature (RC) correction, the standard k-ε model, the RNG k-ε model, the RSM, and LES. These models play a crucial role in capturing the turbulent behavior of fluid flow, particularly in complex systems like cyclone separators, where anisotropic and chaotic turbulence present significant challenges. Additionally, accurately treating the near-wall region, where vorticity and turbulence are most pronounced, is essential for obtaining reliable results. Different approaches, such as wall functions or direct resolution of the viscous sublayer, are employed based on the Reynolds number of the flow.

In conclusion, CFD has revolutionized our understanding of fluid dynamics, equipping engineers with powerful tools to optimize designs, predict system performance, and drive innovation across various industries. As software capabilities and modeling techniques continue to advance, the future of CFD holds promise for even more precise and efficient simulations. This paves the way for groundbreaking discoveries and engineering and scientific research advancements.

References

1. He X, Jiao W, Wang C, Cao W (2019) Influence of surface roughness on the pump performance based on computational fluid dynamics. IEEE Access 7:105331–105341. https://doi.org/10.1109/access.2019.2932021
2. Chu KW, Wang B, Xu DL et al (2011) CFD-DEM simulation of the gas-solid flow in a cyclone separator. Chem Eng Sci 66:834–847. https://doi.org/10.1016/j.ces.2010.11.026
3. Devarampally DR (2017) Development of two-way coupled CFD—DEM model for top spray fluid bed granular using Star CCM +. The State University of New Jersey, New Brunswick
4. Drew DA (1983) Mathematical modeling of two-phase flow. Annu Rev Fluid Mech 15:261–291. https://doi.org/10.1146/annurev.fl.15.010183.001401
5. Fonte CB Jr, Oliveira JAA, De Almeida LC (2015) Dem-Cfd coupling: mathematical modelling and case studies using rocky-Dem® and Ansys Fluent®. In: Eleventh international conference on CFD in the minerals and process industries. CSIRO, Melbourne
6. He Y, Bayly AE, Hassanpour A (2018) Coupling CFD-DEM with dynamic meshing: a new approach for fluid-structure interaction in particle-fluid flows. Powder Technol 325:620–631. https://doi.org/10.1016/j.powtec.2017.11.045

7. Neuwirth J, Antonyuk S, Heinrich S, Jacob M (2013) CFD-DEM study and direct measurement of the granular flow in a rotor granulator. Chem Eng Sci 86:151–163. https://doi.org/10.1016/j.ces.2012.07.005
8. ANSYS (2009) Ansys Fluent 12.0 theory guide, turbulence. ANSYS, Canonsburg
9. Versteeg HK, Malalasekera W (1995) An introduction to computational fluid dynamics: the finite volume method, fluid flow handbook, 2nd edn. Wiley, Hoboken
10. Launder BE, Spalding DB (1974) The numerical computation of turbulent flows. Comput Methods Appl Mech Eng 3:269–289. https://doi.org/10.1016/0045-7825(74)90029-2
11. Xiang J, McGlinchey D, Latham JP (2010) An investigation of segregation and mixing in dense phase pneumatic conveying. Granul Matter 12:345–355. https://doi.org/10.1007/s10035-010-0171-1
12. Cheng XR, Li RN, Gao Y, Guo WL (2013) Numerical research on the effects of impeller pump-out vanes on axial force in a solid-liquid screw centrifugal pump. In: 6th international conference on pumps and fans with compressors and wind turbines-IOP Conference Series: Materials Science and Engineering. Tsinghua University, Beijing
13. ANSYS Fluent Inc (2009) Turbulence. In: ANSYS Inc., ANSYS Europe Ltd (ed) Ansys Fluent 12.0 theory guide, 12th edn. ANSYS, Inc, Canonsburg, pp 1–58
14. Europe A (2021) ANSYS FLUENT user' s guide. ANSYS, Inc, Canonsburg
15. Kock F, Herwig H (2004) Local entropy production in turbulent shear flows: a high-Reynolds number model with wall functions. Int J Heat Mass Transf 47:2205–2215. https://doi.org/10.1016/j.ijheatmasstransfer.2003.11.025
16. Wray TJ, Agarwal RK (2015) Low-Reynolds-number one-equation turbulence model based on k-ε closure. AIAA J 53:2216–2227. https://doi.org/10.2514/1.J053632
17. Spalart P, Allmaras S (1992) A one-equation turbulence model for aerodynamic flows. In: 30th aerospace sciences meeting and exhibition, Reno, NV, USA, pp 5–21
18. Menter FR (1994) Two-equation eddy-viscosity turbulence models for engineering applications. AIAA J 32. https://doi.org/10.2514/3.12149
19. Zhang X, Agarwal RK (2018) Development of various rotation and curvature corrections for eddy-viscosity turbulence models. In: AIAA aerospace sciences meeting, 2018, Kissimmee, FL, USA, pp 1–27
20. Launder BE (1972) Lectures in mathematical models of turbulence. Academic Press, New York, London
21. De AE, Spogis N, Taranto OP, Silva MA (2019) Theoretical study of pneumatic separation of sugarcane bagasse particles. Biomass Bioenergy 127:105256. https://doi.org/10.1016/j.biombioe.2019.105256
22. Boysan F, Ayers WH, Swithenbank J (1982) A fundamental mathematical modeling approach to cyclone design. Trans Inst Chem Eng 60:222–230
23. Narasimha M, Brennan MS, Holtham PN (2006) Numerical simulation of magnetite segregation in a dense medium cyclone. Miner Eng 19:1034–1047. https://doi.org/10.1016/j.mineng.2006.03.013
24. Cortés C, Gil A (2007) Modeling the gas and particle flow inside cyclone separators. Prog Energy Combust Sci 33:409–452. https://doi.org/10.1016/j.pecs.2007.02.001
25. Orszag SA, Yakhot V, Flannery WS, Boysan F (1993) Renormalization group modeling and turbulence simulations. In: International conference, near-wall turbulent flows. Elsevier, Tempe, p 1031
26. Azadi M, Azadi M, Mohebbi A (2010) A CFD study of the effect of cyclone size on its performance parameters. J Hazard Mater 182:835–841. https://doi.org/10.1016/j.jhazmat.2010.06.115
27. Soe TM, Khaing SY (2017) Comparison of turbulence models for computational fluid dynamics simulation of wind flow on cluster of buildings in Mandalay. Int J Sci Res Publ 7:337–350
28. Gimbun J, Chuah TG, Fakhru'l-Razi A, Choong TSY (2005) The influence of temperature and inlet velocity on cyclone pressure drop: a CFD study. Chem Eng Process 44:7–12. https://doi.org/10.1016/j.cep.2004.03.005

29. Griffiths WD, Boysan F (1996) Computational fluid dynamics (CFD) and empirical modelling of the performance of a number of cyclone samplers. J Aerosol Sci 27:281–304. https://doi.org/10.1016/0021-8502(95)00549-8
30. Ma L, Ingham DB, Wen X (2000) Numerical modelling of the fluid and particle penetration through small sampling cyclones. J Aerosol Sci 31:1097–1119. https://doi.org/10.1016/S0021-8502(00)00016-1
31. Gibson MM, Launder BE (1978) Ground effects on pressure fluctuations in the atmospheric boundary layer. J Fluid Mech 86:491–511. https://doi.org/10.1017/S0022112078001251
32. Launder BE, Reece GJ, Rodi W (1975) Progress in the development of a Reynolds-stress turbulence closure. J Fluid Mech 68:537–566. https://doi.org/10.1017/s0022112075001814
33. Wan G, Sun G, Xue X, Shi M (2008) Solids concentration simulation of different size particles in a cyclone separator. Powder Technol 183:94–104. https://doi.org/10.1016/j.powtec.2007.11.019
34. Xiang RB, Lee KW (2005) Numerical study of flow field in cyclones of different height. Chem Eng Process 44:877–883. https://doi.org/10.1016/j.cep.2004.09.006
35. Gronald G, Derksen JJ (2011) Simulating turbulent swirling flow in a gas cyclone: a comparison of various modeling approaches. Powder Technol 205:160–171. https://doi.org/10.1016/j.powtec.2010.09.007
36. Parvaz F, Hosseini SH, Elsayed K, Ahmadi G (2020) Influence of the dipleg shape on the performance of gas cyclones. Sep Purif Technol 233:116000. https://doi.org/10.1016/j.seppur.2019.116000
37. Ansys FI (2001) Modeling turbulence. In: Ansys I, Ansys Europe L (eds) Ansys Fluent guide, 12th edn. Ansys, Inc, Canonsburg, pp 1–102
38. Houben JJH, Weiss C, Brunnmair E, Pirker S (2016) CFD simulations of pressure drop and velocity field in a cyclone separator with central vortex stabilization rod. J Appl Fluid Mech 9:487–499
39. Talbi K (2011) An experimental study and a numerical simulation of the turbulent flow under the vortex finder of a cyclone separator. J Appl Fluid Mech 4:69–75. https://doi.org/10.36884/JAFM.4.01.11903
40. Rotta JC (1962) Turbulent boundary layers in incompressible flow. Prog Aerosp Sci 2:1–95. https://doi.org/10.1016/0376-0421(62)90014-3
41. Martignoni WP, Bernardo S, Quintani CL (2007) Evaluation of cyclone geometry and its influence on performance parameters by computational fluid dynamics (CFD). Braz J Chem Eng 24:83–94. https://doi.org/10.1590/S0104-66322007000100008
42. Shukla SK, Shukla P, Ghosh P (2013) The effect of modeling of velocity fluctuations on prediction of collection efficiency of cyclone separators. Appl Math Model 37:5774–5789. https://doi.org/10.1016/j.apm.2012.11.019
43. Wargadalam VJ, Andira MA, Putra CA et al (2019) Numerical simulation of a small-scale cyclone separator using MP-PIC method. IOP Conf Ser Mater Sci Eng 694:012015
44. Razmi H, Soltani Goharrizi A, Mohebbi A (2019) CFD simulation of an industrial hydrocyclone based on multiphase particle in cell (MPPIC) method. Sep Purif Technol 209:851–862. https://doi.org/10.1016/j.seppur.2018.06.073
45. Spalart PR, Deck S, Shur ML et al (2006) A new version of detached-eddy simulation, resistant to ambiguous grid densities. Theor Comput Fluid Dyn 20:181–195. https://doi.org/10.1007/s00162-006-0015-0
46. Zhang X, Agarwal RK (2019) Applications of a DES model based on Wray-Agarwal turbulence model with rotation and curvature correction. In: AIAA scitech 2019 forum, San Diego, CA, pp 1–16
47. Wilcox DC (2006) Turbulence modeling for CFD, 3rd edn. DCW Industries, La Cañada

Chapter 5
Discrete Element Method (DEM)

Abstract The integration of the simulations of the discrete element method (DEM) with the aerodynamic separation systems represents an important breakthrough for accurate calculation of the particle dynamics in fluid systems in many industrial processes. This chapter describes the advancements associated with utilizing the DEM in aerodynamic separators, shedding light on the fundamental principles and computational techniques that drive this synergistic relationship. As industries such as food and agricultural processing continue to seek more efficient and precise particle separation methods, incorporation of DEM in CFD codes for simulations of particles in fluids emerges as a pivotal solution methodology. This chapter highlights the benefits of employing the DEM, from improving the process efficiency to optimizing the equipment design. Additionally, the chapter explores the key challenges and prospects in this evolving field, providing a roadmap for researchers, engineers, and practitioners to harness the full potential of DEM in aerodynamic systems.

Keywords Particle dynamics · Interaction forces · Modeling approaches · Contact models

5.1 Background of Discrete Element Method (DEM)

The discrete element method (DEM) is a numerical modeling technique first introduced by Cundall and Strack [1]. Its purpose is to complement the Finite Element Method by providing a means of solving complex engineering problems and investigating the natural phenomena of granular flow involving particles of varying shapes and behavior [2]. DEM simulations offer dynamic information that is very difficult or impossible to obtain through formal experimentation, such as individual particle trajectories, transient forces, and analysis of multiple, interacting, deformable, discontinuous, or fractured elements undergoing rotations and large displacements, all extraordinarily challenging or impossible to obtain through formal experimentation [3, 4]. With the advent of advanced computers, simulation software can now iterate calculations millions of times per second for each element in the

L. Zhou et al., *Discrete Element Method for Multiphase Flows with Biogenic Particles*, https://doi.org/10.1007/978-3-031-67729-8_5

system. As a result, the practicality of DEM has significantly increased, enabling simulations of processes involving over a million particles [5]. In DEM modeling, the interaction among the particles is treated as a dynamic process and is solved using an explicit numerical scheme. Each stressed particle's contact forces, rotations, momentum, and displacements are tracked by describing each contact of the individual particles [6]. The primary assumption in DEM is that each discrete element has distinct boundaries that physically separate it from every other element in the analysis.

Particle movement within a domain can be a complex and intricate process. To accurately model this movement, translating and rotating behavior equations must be applied to each particle, with particles making contact through boundary forces. As a result of these interactions, deformation can occur, decoupled from the mean motion by expressing it as the sum of the bodies' normal modes. These newly derived decoupled modal equations can then be solved using an explicit central difference scheme, with the final solution obtained through modal superposition [4, 7]. In recent years, the use of the DEM has become increasingly popular in the fields of harvest and postharvest processing of grain, as well as in various other industries such as granular mixing, food drying, pharmaceuticals, mining, and agricultural processing. This is due to the compact and accurate characterization of actual conditions that DEM offers, allowing for the prediction of various processes with great precision [8]. Unlike mining and chemical industry, however, DEM is not being widely applied in agricultural processes because of various particle property issues arising from the biological origins of grain and food products.

In the agricultural industry, DEM in agricultural and food processing applications has been considerably developed to describe the flow in chutes and dryers [9–12], the aerodynamic conveying systems [13, 14], the discharging silo [15], the transportation and compression of seeds [16], the dynamics discharge of particles in hoppers [17], the shaking separation of paddy and brown rice [18], the grain–straw separation screening in combine harvester [19, 20], the fluidization and spray drying of food powders [21], the grain transportation and aerodynamic separation [22], the grains segregation and seed motion in seed feeding device [23, 24], the screening separation of crop seeds [25, 26], and the mixing process in a bucket elevator [27]. DEM has also been helpful in the modeling of interactions between soft materials, such as fruits like grapes [28], apples [29], or biomass [30]. A complete description and valuable review of the DEM can be found in previous literature [4, 7, 31–33].

5.2 Approaches in DEM Modeling

Granular dynamics modeling using DEM techniques is a fascinating field of study that involves simulating the behavior of particles during contact or collision [34]. Hard-sphere and soft-sphere modeling are two popular approaches for modeling granular dynamics using DEM. When modeling granular dynamics using the

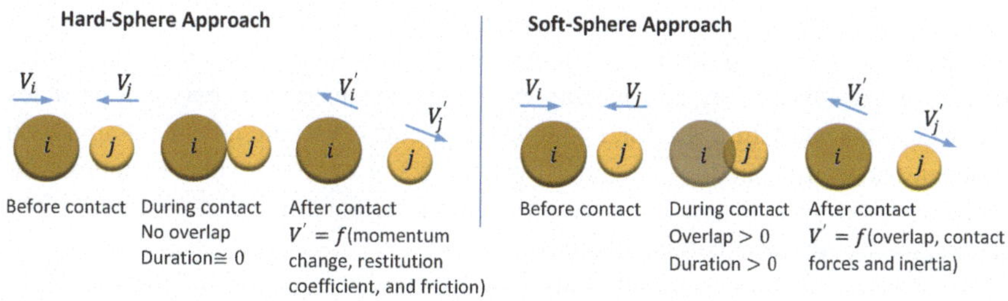

Fig. 5.1 Illustration of hard-sphere and soft-sphere approaches [35, 36]

hard-sphere approach, particles are treated as non-deformable spheres. This simplifies collision modeling by assuming no interpenetration or deformation during particle collisions. On the other hand, soft-sphere modeling involves considering the deformation of spheres during collisions and the duration of contact between the particles. This approach provides a more detailed understanding of the dynamics of granular materials. To better understand the differences between these approaches, see Fig. 5.1, which schematically illustrates the critical distinctions between hard-sphere and soft-sphere modeling [35, 36].

In the fascinating world of granular materials, the hard-sphere modeling approach has gained immense popularity for efficiently handling overlapping and deformation between particles during contact [37]. However, to simplify the modeling process, this approach assumes the neglect of these effects, giving rise to a non-smooth DEM [38]. In such cases, the motion of particles and energy loss during a collision can be predicted using the shock laws and coefficients of restitution. In this approach, collisions are assumed to occur quickly, limiting the interaction between the two particles and ignoring multiple contacts [39]. This results in different time steps, depending on the time between each collision. Therefore, it is well-suited for rapid granular flow simulations, especially when the particulate material is partially or wholly fluidized [36]. By understanding the mechanics of this modeling approach, it is possible to better comprehend the behavior of granular material, paving the way for innovative solutions in diverse fields such as agriculture, mining, and civil engineering.

In contrast to the soft-sphere approach, a more advanced approach allows for overlapping at contact points and considers the strong contact forces between the particles [3, 35]. To successfully employ this approach, it is crucial to accurately model the elastic restitution and friction during particle collisions [40]. The soft-sphere approach relies on iterative schemes, making it highly sensitive to the time step. Therefore, to ensure stability during simulation, a constant time step interval less than the duration of the contact is necessary [37, 41]. In many soft-sphere models, the duration of the contact is artificially increased by allowing for softer interactions, which reduces the required CPU time. However, this may not accurately represent the deformation at the contacts between biogenic particles. By contrast, the advanced approach utilizes the deformations caused by collisions to calculate elastic, plastic, and frictional forces between particles [4]. This makes the advanced approach particularly useful for particulate mechanics models of agricultural cereals and crops. Overall, by considering the complex interactions between particles and accurately modeling the forces involved in collisions, the advanced approach represents a significant improvement over the traditional soft-sphere method. Its applications in agricultural research demonstrate its potential to revolutionize our understanding of particulate mechanics.

By utilizing the numerical integration of Newtonian equations of motion, these models can elucidate the complex dynamics of particles interacting with each other and external forces. One of the most significant factors contributing to the accuracy of soft-sphere models is the interparticle forces that govern the system's behavior [42, 43]. Notably, the magnitude of the particle overlap is a critical factor in determining the interparticle contact force, as discussed in detail below. An essential advantage of soft-sphere models is their ability to simulate the behavior of dense-phase bulk granular materials [44], where multiple particle contacts play a vital role in the overall dynamics of the system, particularly in quasi-static systems [3]. These models start by solving the equations that govern the particles' linear translation and rotational motion [36].

Soft-sphere models effectively study complex systems as they accurately account for interparticle forces and multiple particle contacts, providing a robust approach to understanding granular materials' behavior. They have the potential to reveal the underlying mechanisms governing particle behavior, making them a significant step forward in this field [37]. However, hard-sphere models are computationally inexpensive, fast, and preferred for non-dense flow. While they help provide a limited description of the dense material's response involving multiple simultaneous contacts, they are not as effective as soft-sphere approaches in describing bulk material physics [45]. Therefore, this work focuses on soft-sphere algorithms, commonly used in the grain and food processing industries, and does not consider hard-sphere models further.

5.3 Governing Equations of Motion

Soft-sphere DEM modeling tracks the motion of each particle by calculating the contact forces and displacements using an explicit numerical scheme and a small time step [1]. The particle motion can be separated into two parts, translation motion calculated using Newton's second law and rotational motion described by Euler's law as shown in Eqs. (5.1) and (5.2), respectively [46–48]. Interstitial fluid, typically air in aerodynamic systems, occupies the spaces between the particles. When the interstitial fluid forces the particles, its effect must be considered. However, if the particles are large and heavy, they are unaffected by the interstitial fluid [44].

$$m_i \frac{d^2}{dT^2} x_i = \sum_{j=1}^{n_i^c} F_{ij}^c + F_i^g + F_i^f + \sum_{k=1}^{n_i^c} F_{ik}^{nc} \tag{5.1}$$

$$I_i \frac{d^2}{dT^2} \Phi_i = \sum_{j=1}^{n_i^c} M_{ij} \tag{5.2}$$

where m_i is the particle mass; x_i and Φ_i are the position and the orientation in the space of particle i, respectively; F_{ij}^C and M_{ij} are the total contact force and momentum acting on particle i by particle j or wall, respectively; f_i^g is the gravitational force; F_i^f is the fluid-particle interaction force on the particle i; F_{ik}^{nc} is a term of non-contact force acting on particle i by particle k or other sources; I_i is the moment of inertia of particle i; and T is the time.

In soft-sphere DEM modeling, the contact between two particles does not occur at a single point but in a specific area, which can cause slight overlapping of the particles on the normal and tangential coordinates (n & t) [3]. Ignoring particle overlapping can significantly affect the simulation results [49]. Therefore, the total contact force acting on the particle at the contact plane must be analyzed in n & t coordinates, which can be calculated mathematically using Eq. (5.3). Figure 5.2 provides a schematic illustration of the typical contact force vectors with overlapping and torques involved in soft-sphere DEM modeling.

$$F_{ij}^C = F_{ij}^n + F_{ij}^t \tag{5.3}$$

where F_{ij}^n and F_{ij}^t are the normal and tangential contact forces on the particle i by contacting particle j, respectively.

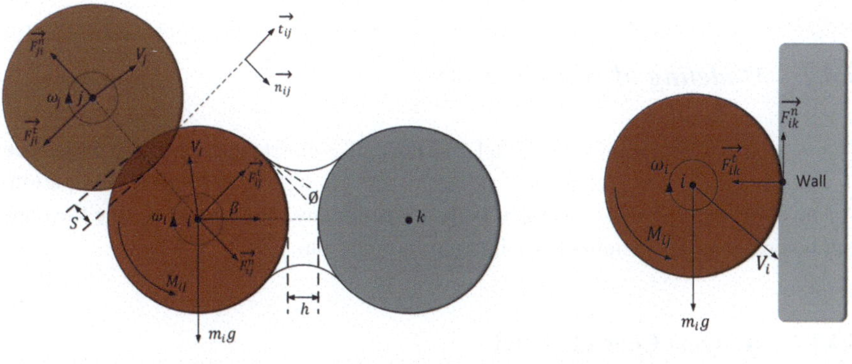

Fig. 5.2 Contact forces analysis during the particle collision with overlapping [49]

5.4 Modeling of Contact Force

In granular flows, interparticle contacts generate the primary forces, which can be mechanically described by how the particles perceive each other. The DEM adopts a philosophy that employs rigid springs acting normally and tangentially to the contact plane to calculate interparticle forces. Although DEM simulations efficiently predict numerous processes in the agricultural industry, modeling contact models requires a comprehensive understanding of particle interactions under varying loads. Precise values of physical and mechanical parameters of granular materials with biological origins are critical for contact model development in DEM [50]. The high moisture content of these particles makes them highly deformable during contact, necessitating the generation of different contact models based on elastic, viscous, plastic, dry friction, and adhesive interactions [33].

Theoretical contact models are typically based on the assumptions of isotropic and elastic particles with completely smooth contacts. However, agricultural grains, which come in various geometries and shapes, challenge these assumptions. Most are convex, and none are purely elastic bodies [33]. As a result, developing effective contact models for agricultural materials has limitations. In any DEM, the contact forces consist of normal and tangent forces. For spherical particles, the contact plane is perpendicular to the line connecting the two spheres' centers. However, for non-round particles, determining the contact plane can be more complex. It may involve identifying the closest points between the two particles, a particle and a triangle, or the points with maximum distance overlap in physical contact [51]. This chapter delves into the common contact models researchers use for agricultural cereal grains and seeds, including interaction properties such as coefficients of restitution, static and rolling friction, plastic or viscous damping, and adhesion. Understanding these models can improve the understanding of the behavior of these complex materials and enhance our ability to design effective agricultural processes.

5.4.1 Modeling of Normal Force

The primary requirements to model the normal contact force are that the force should be repulsive and that the model must allow significant energy dissipation. For this purpose, several investigators developed many models to describe the contact behavior of the systems with the granular medium.

5.4.1.1 Hertzian Contact Model

In the illustrious footsteps of Hertz from the nineteenth century and based on the principles of elasticity [52, 53], the normal force of contact is ingeniously modeled by analyzing the response of two elastic spheres in contact over a small circular area. This exceptional model has been extensively utilized to determine the modulus of elasticity for various agricultural grains with a convex shape [54]. The model is often called the Hertzian spring dashpot model, wherein the normal contact force is expressed as the sum of elastic and damping forces. This mathematical model can be elegantly expressed incrementally by accounting for the overlap during contact, as demonstrated in Eqs. (5.4), (5.5), and (5.6).

$$F_{ij}^{n} = \frac{4}{3}E^{*}\sqrt{R^{*}S^{n^{3}}} + \eta\left(\frac{4}{3}m^{*}E^{*}\sqrt{R^{*}S^{n}}\right)^{1/2}\frac{d}{dT}S^{n} \tag{5.4}$$

$$\frac{1}{E^{*}} = \frac{\left(1-v_i^2\right)}{E_i} + \frac{\left(1-v_j^2\right)}{E_j} \tag{5.5}$$

$$R^{*} = \frac{R_i R_j}{R_i + R_j} \tag{5.6}$$

where E^* and R^* are equivalent to Young's modulus and radius; E_i, E_j and ν_i, ν_j are Young's moduli and Poisson's ratios of the two contacting spheres of radii R_i, R_j, respectively; S^n is the normal overlapping; η is the damping ratio, which is a dimensionless parameter related to the restitution coefficient that can be calculated as expressed in Eq. (5.7) [55, 56].

$$\eta = -\ln\varepsilon\sqrt{\frac{5}{\ln^2\varepsilon + \pi^2}} \tag{5.7}$$

where ε is the coefficient of restitution defined by the DEM user according to the study case, and m^* is the equivalent mass for the particles in contact that can be calculated according to Eq. (5.8).

$$\frac{1}{m^{*}} = \frac{1}{m_i} + \frac{1}{m_j} \tag{5.8}$$

5.4.1.2 Elastic-Plastic Contact Mode

The hysteretic linear spring model, initially proposed by Walton and Braun [57], has proven to be a game changer in the elastic-plastic or repulsive-dissipative normal contact simulation field. This model allows for the accurate simulation of plastic energy dissipation on contact without compromising simulation time, which can be a critical factor in the research. The energy dissipation is independent of neighboring particles' velocities and loading rates, making it insensitive to other contacts, and the coefficient of restitution is velocity-independent. This makes the model ideal for accurately modeling compressible materials by stimulating the flow of non-adhesive particles with precision [59–61]. The model expressed mathematically through Eqs. (5.9), (5.10), (5.11), (5.12), (5.13), and (5.14) provides a comprehensive and descriptive framework for the linear hysteresis model used in the DEM method [51, 58, 62–64]. This model can achieve accurate and efficient simulations, paving the way for groundbreaking discoveries in contact mechanics.

$$F_i^{n,T} = \begin{cases} \min\left(F_i^{n,(T-\Delta T)} + K_u^n \Delta S^n, K_l^n S^{n,T}\right), & \text{if} \Delta S^n = 0 \\ \max\left(F_i^{n,(T-\Delta T)} + K_u^n \Delta S^n, 0.001 K_l^n S^{n,T}\right), & \text{if} \Delta S^n < 0 \end{cases} \tag{5.9}$$

$$\Delta S^n = S^{n,T} - S^{n,T-\Delta T} \tag{5.10}$$

where $F_i^{n,T}$ and $F_i^{n,(T-\Delta T)}$ are the normal elastic-plastic contact forces acting on the particle i at the current time T and at the previous time $T - \Delta T$, respectively, where ΔT is the timestep, ΔS^n is the change in the normal contact overlap during the time step; $S^{n,T}$ and $S^{n,T-\Delta T}$ are the normal overlap values at the current and at the previous time, respectively; and K_u^n and K_l^n are the unloading and loading contact stiffnesses, respectively.

Note that the unloading force is limited by the value of $0.001K_l^n S^{n,T}$ to ensure that the normal force becomes zero with zero overlappings. As shown in Fig. 5.3, a schematic sketch can demonstrate the loading and unloading response's overlapping circle (A–C) due to the normal force effect, where the hatched area is

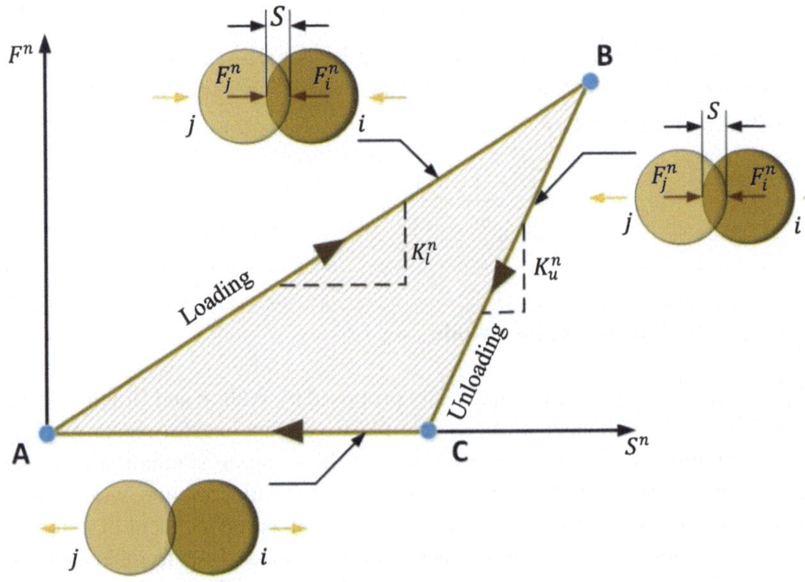

Fig. 5.3 Normal force-overlapping response for the hysteretic linear spring model

represented by energy dissipation during the particle contact. Besides, K_l^n and K_u^n are the slope of loading and unloading lines, respectively, and the plastic deformation only happens during the contact of the particles *i* and *j*; after that, the residual deformation is neglected.

The coefficient of restitution can be determined by the user, which represents the energy dissipation due to particle–particle or particle–wall contact, as well as the loading and unloading stiffness represented in Eqs. (5.11) and (5.12) can be considered by considering the value of Young's modulus and the particle geometry.

$$\frac{1}{K_l^n} = \begin{cases} \frac{1}{K_{li}^n} + \frac{1}{K_{lj}^n} & \text{Particle – Particle} \\ \frac{1}{K_{li}^n} + \frac{1}{K_{lw}^n} & \text{Particle – wall} \end{cases} \tag{5.11}$$

$$K_u^n = \frac{K_l^n}{\varepsilon^2} \tag{5.12}$$

$$K_{li}^n = E_i Z \tag{5.13}$$

$$K_{lk}^n = E_w Z \tag{5.14}$$

where E_i and E_w are Young's modulus of particle and boundary, respectively, and Z represents the particle size or geometry. This model should be applied to adequately select exceptional cases of the behavior of dry and hard seeds with elastic behavior in the normal direction, such as the application of discharging silo with rapeseeds and the aerodynamic systems for seeds and foreign materials separation [15].

5.4.1.3 Visco-Elastic Contact Model

The simple model to describe the visco-elastic behavior of the particles is called a linear elastic or viscous model. The advantage of this model is that it takes less time and machine resources to process but provides less accurate results than the hysteretic linear spring model [51]. When visco-elastic particles are in contact, the normal contact force is given by the sum of the elastic repulsive force and a damping force proportional to the velocity, as formulated in Eq. (5.15) [33, 65].

$$F_i^n = K_l^n S^n + C^n \frac{d}{dt} S^n \tag{5.15}$$

where C^n is the normal damping coefficient and is defined as follows in Eq. (5.16).

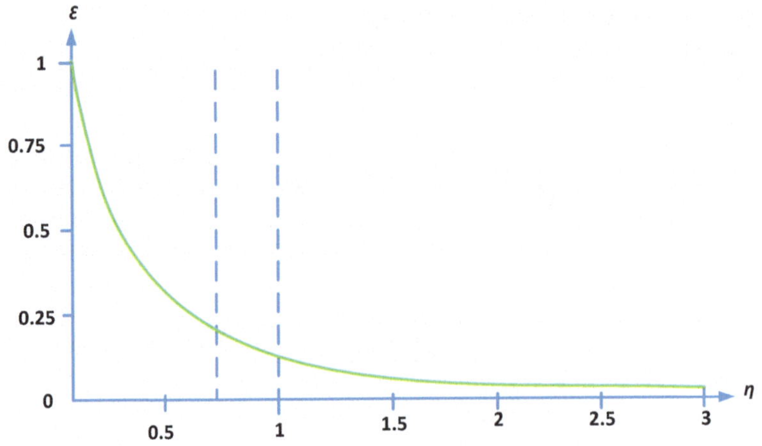

Fig. 5.4 The relationship between the damping ratio and the restitution coefficient [66]

$$C^n = 2\eta\sqrt{m^* K_l^n} \tag{5.16}$$

The relationship between the damping ratio and the restitution coefficient can be described in Eq. (5.20) [66]. Figure 5.4 describes in detail the relationship between the damping ratio and the restitution coefficient given by Eq. (5.17). The graph shows that the increase in the damping ratio and the decrease in the restitution coefficient are due to the inverse relationship between the two parameters.

$$\varepsilon = \begin{cases} \exp\left[-\dfrac{\eta}{\sqrt{1-\eta^2}}\left(\pi - \arctan\dfrac{2\eta\sqrt{1-\eta^2}}{1-2\eta^2}\right)\right] & \text{if } 0 \le \eta < \dfrac{1}{\sqrt{2}} \\ \exp\left[-\dfrac{\eta}{\sqrt{1-\eta^2}}\left(\arctan\dfrac{2\eta\sqrt{1-\eta^2}}{1-2\eta^2}\right)\right] & \text{if } \dfrac{1}{\sqrt{2}} \le \eta \le 1 \\ \exp\left[-\dfrac{\eta}{\sqrt{\eta^2-1}}\left(\ln\dfrac{2\eta\sqrt{1-\eta^2}}{1-2\eta^2}\right)\right] & \text{if } \eta > 1 \end{cases} \tag{5.17}$$

For soft biological materials such as apples, potatoes, and tomatoes, in addition to the soft and high moisture content of seeds, the visco-elastic contact model is efficient and appropriate for predicting the energy dissipation and the impact reaction [67, 68]. For dry and hard seeds such as rapeseed [15], the elastic-plastic contact model works well to estimate the contact and impact behavior.

5.4.2 Modeling of Tangential Force

5.4.2.1 Linear Spring Coulomb Model

The linear elastic-frictional model, aptly named for its ability to exhibit elastic behavior before the commencement of friction, is a remarkable model that can effectively regenerate Coulomb frictional contact behavior for both the static and dynamic frictional values. In determining the total tangential contact force during tangential contact, the sliding of the particles plays a crucial role. This effect is expressed by Eq. (5.18), which has been documented in various research papers [59, 62, 64].

$$F_i^{t,T} = \min\left(F_i^{t,(T-\Delta T)} + K_l^n \Delta S^t, \mu F_i^{n,T}\right) \tag{5.18}$$

where $F_i^{t,T}$ and $F_i^{t,(T-\Delta T)}$ are the tangential contact forces at the current and previous time of the simulation, respectively ; ΔS^t is the change in the tangential overlapping during the time step; and μ is the friction coefficient, which can be defined as a static and dynamic coefficients (μ_s , μ_d) depending on the tangential sliding contact, as given in Eq. (5.19) [62, 63].

$$\mu = \begin{cases} \mu_s & \text{if no sliding at the contact} \\ \mu_d & \text{if sliding at the contact} \end{cases} \tag{5.19}$$

It is worth noting that when the tangential force surpasses the limit of $\mu_s F_i^{n,T}$, the sliding of particles takes place on the contact surface. On the other hand, when this force falls below this value, the contact is considered non-sliding once again. The effectiveness of this model has been proven in the vertical pneumatic separation of sugarcane bagasse particles [62]. In conclusion, the linear elastic-frictional model is highly sophisticated and exhibits both elastic and Coulomb frictional contact behavior, making it an ideal model for various industrial applications. Its ability to accurately simulate sliding particle contact dynamics makes it a valuable tool in material science and engineering.

5.4.2.2 Mindlin-Deresiewicz Model

The elastic-frictional contact model (Mindlin-Deresiewicz Model) has been widely adopted due to its ability to accurately predict the behavior of elastic spheres under varying oblique forces [69]. This model comprises loading histories with varying normal and tangential forces and a minimal range of tangential displacement between the spheres in contact [70, 71]. The tangential force in this model is defined by Eq. (5.20), which ensures high precision and reliability [51]:

$$F_i^{t,T} = \mu F_i^n \left[1 - \left(1 - \frac{\min\left(\left|s^t\right|, s_{\max}^t\right)}{s_{\max}^t} \right)^{3/2} \right] \frac{s^t}{\left|s^t\right|}$$

$$+\eta^t \left(\frac{6\mu\, m^* F_i^n}{s_{\max}^t} \sqrt{1 - \left(\frac{\left|s^t\right|}{s_{\max}^t} \right)} \right)^{1/2} \left(\widehat{n_c} \times \frac{d}{dT} S^t \right) \times \widehat{n_c} \tag{5.20}$$

where s^t is the tangential relative displacement at the contact; $\widehat{n_c}$ is the unit vector in the normal direction at the contact; η^t is the tangential damping ratio, which is estimated through Eq. (5.21); $s_{\max}^t$ is the maximum relative tangential displacement at which particles begin to slide that can be formulated in Eq. (5.22); and $\frac{d}{dT} S^t$ is the relative tangential velocity at the contact.

$$\eta^t = -\ln \varepsilon \frac{1}{\sqrt{\ln^2 \varepsilon + \pi^2}} \tag{5.21}$$

$$s_{\max}^t = \mu \left(\frac{1 - v_i}{2 - v_i} + \frac{1 - v_j}{2 - v_j} \right)^{-1} S^n \tag{5.22}$$

However, it is noteworthy that the Mindlin-Deresiewicz model does not function optimally when applied to non-spherical particles with rolling resistance. Therefore, the rolling resistance must be omitted when utilizing this model with non-round particles [51]. Nevertheless, to enhance the model's capabilities, researchers have successfully adapted it to simulate the flow of soybeans down an inclined chute [11]. Moreover, the frictional contact model of spherical particles was also developed based on the Mindlin-Deresiewicz model to accurately predict the particle–particle slip and the sliding behavior of stainless steel beads, peas, and rapeseed [72]. By constantly improving and expanding on this model, researchers can gain a deeper understanding of the physics of particle interactions and improve their ability to predict and control the behavior of granular materials.

5.4.3 *Stiffness Contact Model*

The Hertz-Mindlin contact model is widely used in DEM simulations as a primary law to calculate particle interaction forces in the tangent directions [55, 69, 71]. It is also known as a linear spring-dashpot model, which includes springs, dashpots, and frictional sliders to represent the system's stiffness, as shown in Fig. 5.5 [73]. The model is based on Hertz's theory [53] for normal stiffness and Mindlin and Deresiewicz's theory [69] for tangential stiffness. Both are calculated using elastic

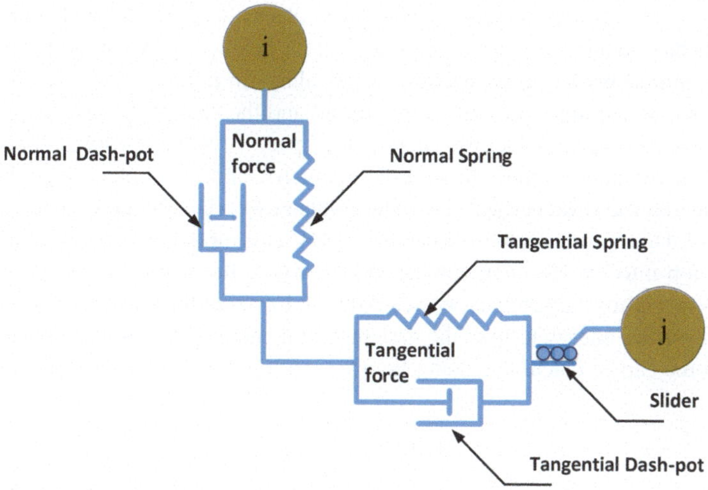

Fig. 5.5 Hertz-Mindlin contact model with viscous damping and frictional slider [73]

springs and dashpots in their respective directions to describe normal stiffness and Mindlin and Deresiewicz's theory to describe tangential stiffness. Young's modulus parameter, friction coefficient, and coefficient of restitution are used to express spring stiffness, dashpot coefficient, and friction sliders, respectively [58, 74, 75].

The Hertz-Mindlin contact model has been extensively applied to simulate the interaction of agricultural particulate matter such as seeds and grains. Calibration and validation studies have shown that this model is highly reliable and accurately predicts the behavior of particles. For instance, it has been used to simulate the motion of rice grain and short straw in an air-and-screen cleaning device [75], the flow of corn-shaped particles in a spouted bed [76], and the motion of rapeseed and wheat grain in a seed feeding device [24]. The model has also been used to predict the flow behavior of corn grain in a commercial screw auger [77], simulate the motion of rice grain and rice straw in a variable-amplitude screen box [19], and optimize the air-blowing maize precision seed-metering device [78]. Overall, the Hertz-Mindlin contact model is a powerful tool for simulating and predicting the behavior of agricultural particulate matter. Its high reliability and accuracy make it an ideal choice for various applications in this field.

5.4.4 Adhesion Contact Model

Intimate contact between biological materials can create a sticky situation that requires mechanical force to separate. Such adhesive forces can lead to powders and small particle clumping, causing storage and handling problems [79, 80]. Numerical simulations must incorporate repulsive and attractive forces to model this behavior accurately [51]. Moisture content, van der Waals, electrostatic forces, and liquid bridges can contribute to adhesion between biological particles [81]. When modeling larger particles, liquid bridge force becomes particularly important [65, 82]. This force arises from the capillary pressure of the liquid and the sum of its surface tension. It is influenced by various factors, such as the space between particles, the liquid's surface tension and volume, the particle geometry, and the contact angle between the liquid and the solid surface [83]. The Johnson-Kendall-Roberts (JKR) model is widely used to represent adhesive behavior [84]. This model incorporates the effects of adhesion into the Hertzian contact model, where the contact area between two particles is slightly larger than predicted by the Hertzian theory [51]. This is due to adding a surface energy term to the system's total energy. The normal JKR adhesive model force can be calculated using Eq. (5.23) [51]:

$$F_{i_{\mathrm{adh}}}^{n,T} = \frac{E^* a^3}{6R^*} \tag{5.23}$$

where $F_{i_{\mathrm{adh}}}^{n,T}$ is the normal adhesive contact force at the current time, and a is the separation distance of contact between particles or between a particle and a boundary, as shown in Fig. 5.6, where it is related to the normal overlap using the expression introduced in Eq. (5.24) [33, 51].

$$S^n = \frac{a^2}{4R^*} - \left(\frac{\pi \Gamma a}{E^*}\right)^{1/2} \tag{5.24}$$

where Γ is the surface energy and can be used to calculate the maximum liquid bridge tensile force that occurs between two particles as formulated in Eq. (5.25) [65].

$$P_{\mathrm{max}} = -2\pi\Gamma E^* \tag{5.25}$$

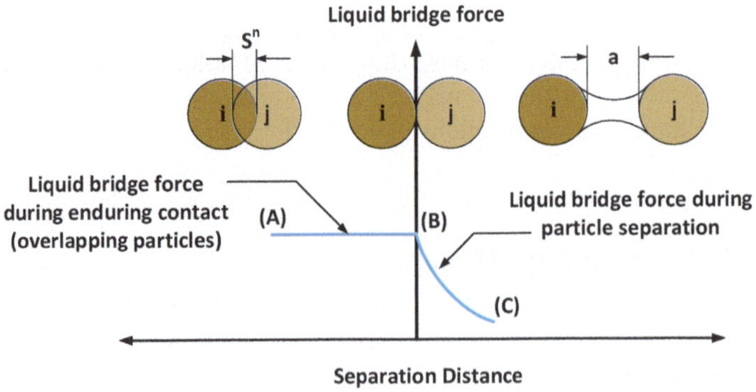

Fig. 5.6 Schematic diagram of the liquid bridge force as a function of the separation distance between the two particles [17]

where P_{max} is the maximum liquid bridge tensile force since the effect of its behavior can be observed through the DEM simulation of the particle's free-flowing and the discharge rate from hoppers of different inclinations [17, 85]. Another study on agricultural material reported that the maximum liquid bridge tensile force between wet rapeseeds is approximately tenfold the seed weight, which means that the cohesive behavior can occur in the case of some particles' of millimeters size [33].

The JKR model is a well-established theoretical framework that finds widespread acceptance in the context of adhesive elastic sphere particles. One of its most significant advantages is that it allows for experimental measurement of surface energy, which makes it ideal for simulating perfect spheres without the need for calibration [51]. On the other hand, when it comes to adhesion between particles resulting from liquid bridge forces, such as the high attraction between polar molecules of water, an alternative adhesive model known as the constant adhesive force model is often used [62, 73]. This model relies on the following essential parameters: the minimum distance between particles or the boundary, the application of adhesive forces, and the force fraction expressed as a function of the particle gravity force. Once the force fraction reaches unity, the adhesive force becomes equivalent to the gravity force applied to the particle. In cases where two particles of different masses make contact, the smaller mass is used in calculating the gravitational force [51], as outlined in Eq. (5.26).

$$F_{i_{\text{adh}}}^{n,T} = \begin{cases} 0 & if - \Delta S^n > \delta_{\text{adh}} \\ f_{\text{adh}} \times g \times \min\left(m_i, m_j\right) & \text{if} - \Delta S^n < \delta_{\text{adh}} \end{cases} \tag{5.26}$$

where f_{adh} and δ_{adh} are the force fraction and the adhesive distance, respectively. Note that, when $\Delta S^n \geq 0$, the particles start adhering to each other, but if $\Delta S^n < 0$, the particles separate [73]. This model is applied effectively to represent the adhesion force of sugarcane bagasse particles during the pneumatic separation [62].

5.5 Summary

The DEM is a potent numerical modeling technique that has revolutionized our capacity to study and comprehend the behavior of granular materials within various industries, including agriculture, mining, and food processing. Initially, DEM complemented the finite element method by providing dynamic insights into granular flow phenomena involving particles of varying shapes and behaviors. The versatility of DEM simulations lies in their capacity to offer detailed information such as individual particle trajectories, transient forces, and the analysis of multiple interacting elements undergoing rotations and significant displacements. These aspects are often challenging or impossible to obtain through traditional experimentation. With the advancement of computer technologies, DEM simulations can now iterate calculations millions of times per second for each element in a system, enabling the simulation of processes involving over a million particles. DEM modeling has two popular approaches, namely, hard-sphere and soft-sphere modeling, each offering distinct advantages depending on material properties and the desired level of detail. The hard-sphere approach simplifies collision modeling by assuming non-deformable particles, making it suitable for rapid granular flow simulations. Conversely, the soft-sphere approach considers particle deformations and strong contact forces, providing a more detailed understanding of granular material dynamics. This advanced approach is precious in modeling agricultural cereals and crops, where particle properties vary significantly. A crucial aspect of DEM modeling is the accurate representation of interparticle forces, including regular and tangential forces during contact. Various contact force models have been developed to describe these interactions, such as the Hertzian, elastic-plastic, visco-elastic, and linear spring Coulomb models. These models enable researchers to predict particle behavior, energy dissipation, and the effects of adhesion between particles, which is essential for understanding the complex dynamics of granular materials.

DEM has found extensive applications in agricultural and food processing, modeling grain flow, aerodynamic conveying systems, seed transportation, mixing processes, and more. By accurately simulating these processes, DEM provides a pathway to optimize agricultural operations, improve equipment design, and enhance product quality. In essence, the DEM remains a valuable tool for researchers and engineers, providing unparalleled insights into the behavior of granular materials across diverse industries.

References

1. Cundall PA, Strack ODL (1979) A discrete numerical model for granular assemblies. Géotechnique 29:47–65. https://doi.org/10.1680/geot.1979.29.1.47
2. Dewicki G (2003) Bulk material handling and processing—numerical techniques and simulation of granular material. Bulk Solids Handl Int J Storing Handl Bulk Mater 23:110–113
3. Zhu HP, Zhou ZY, Yang RY, Yu AB (2007) Discrete particle simulation of particulate systems: theoretical developments. Chem Eng Sci 62:3378–3396. https://doi.org/10.1016/j.ces.2006.12.089
4. Boac JM, Ambrose RPK, Casada ME et al (2014) Applications of discrete element method in modeling of grain postharvest operations. Food Eng Rev 6:128–149. https://doi.org/10.1007/s12393-014-9090-y
5. Berger R, Kloss C, Kohlmeyer A, Pirker S (2015) Hybrid parallelization of the LIGGGHTS open-source DEM code. Powder Technol 278:234–247. https://doi.org/10.1016/j.powtec.2015.03.019
6. Theuerkauf J, Dhodapkar S, Jacob K (2007) Modeling granular flow using discrete element method—from theory to practice. Chem Eng 114:39–47
7. Williams JR, Hocking G, Mustoe G (1985) The theoretical basis of the discrete element method. In: NUMETA '85 conference, Balkema, Rotterdam, pp 897–906
8. Zhu HP, Zhou ZY, Yang RY, Yu AB (2008) Discrete particle simulation of particulate systems: a review of major applications and findings. Chem Eng Sci 63:5728–5770. https://doi.org/10.1016/j.ces.2008.08.006

9. Keppler I, Kocsis L, Oldal I et al (2012) Grain velocity distribution in a mixed flow dryer. Adv Powder Technol 23:824–832. https://doi.org/10.1016/j.apt.2011.11.003
10. Khatchatourian OA, Binelo MO, de Lima RF, De Lima RF (2014) Simulation of soya bean flow in mixed-flow dryers using DEM. Biosyst Eng 123:68–76. https://doi.org/10.1016/j.biosystemseng.2014.05.003
11. Zhang X, Vu-Quoc L (2000) Simulation of chute flow of soybeans using an improved tangential force-displacement model. Mech Mater 32:115–129. https://doi.org/10.1016/S0167-6636(99)00043-5
12. Weigler F, Scaar H, Mellmann J (2012) Investigation of particle and air flows in a mixed-Flow dryer. Dry Technol 30:1730–1741. https://doi.org/10.1080/07373937.2012.703742
13. Sturm M, Wirtz S, Scherer V, Denecke J (2010) Coupled DEM-CFD simulation of pneumatically conveyed granular media. Chem Eng Technol 33:1184–1192. https://doi.org/10.1002/ceat.201000162
14. Lim EWC, Wang CH, Yu AB (2006) Discrete element simulation for pneumatic conveying of granular material. AICHE J 52:496–509. https://doi.org/10.1002/aic.10645
15. Parafiniuk P, Molenda M, Horabik J (2013) Discharge of rapeseeds from a model silo: physical testing and discrete element method simulations. Comput Electron Agric 97:40–46. https://doi.org/10.1016/j.compag.2013.06.008
16. Raji AO, Favier JF (2004) Model for the deformation in agricultural and food particulate materials under bulk compressive loading using discrete element method. II: compression of oilseeds. J Food Eng 64:373–380. https://doi.org/10.1016/j.jfoodeng.2003.11.005
17. Anand A, Curtis JS, Wassgren CR et al (2009) Predicting discharge dynamics of wet cohesive particles from a rectangular hopper using the discrete element method (DEM). Chem Eng Sci 64:5268–5275. https://doi.org/10.1016/j.ces.2009.09.001
18. Sakaguchi E, Suzuki M, Favier JF, Kawakami S (2001) Numerical simulation of the shaking separation of Paddy and Brown rice using the discrete element method. J Agric Eng Res 79:307–315. https://doi.org/10.1006/jaer.2001.0706
19. Ma Z, Li Y, Xu L (2015) Discrete-element method simulation of agricultural particles' motion in variable-amplitude screen box. Comput Electron Agric 118:92–99. https://doi.org/10.1016/j.compag.2015.08.030
20. Lenaerts B, Aertsen T, Tijskens E et al (2014) Simulation of grain-straw separation by discrete element modeling with bendable straw particles. Comput Electron Agric 101:24–33. https://doi.org/10.1016/j.compag.2013.12.002
21. Woo MW, Daud WRW, Mujumdar AS et al (2010) Role of rheological characteristics in amorphous food particle-wall collisions in spray drying. Powder Technol 198:251–257. https://doi.org/10.1016/j.powtec.2009.11.015
22. Adewumi BA (2008) 2D modeling of grain transport and separation in the horizontal air stream. In: Food processing automation conference CD-Rom, pp 1–10
23. Guo Y, Wu CY, Kafui KD, Thornton C (2011) 3D DEM/CFD analysis of size-induced segregation during die filling. Powder Technol 206:177–188. https://doi.org/10.1016/j.powtec.2010.05.029
24. Lei X, Liao Y, Liao Q (2016) Simulation of seed motion in seed feeding device with DEM-CFD coupling approach for rapeseed and wheat. Comput Electron Agric 131:29–39. https://doi.org/10.1016/j.compag.2016.11.006
25. Li J, Webb C, Pandiella SS, Campbell GM (2002) A numerical simulation of separation of crop seeds by screening – effect of particle bed depth. Food Bioprod Process Trans Inst Chem Eng Part C 80:109–117. https://doi.org/10.1205/09603080252938744
26. Li J, Webb C, Pandiella SS, Campbell GM (2003) Discrete particle motion on sieves – a numerical study using the DEM simulation. Powder Technol 133:190–202. https://doi.org/10.1016/S0032-5910(03)00092-5
27. Boac JM, Casada ME, Maghirang RG, Harner JP (2012) 3-D and quasi-2-D discrete element modeling of grain commingling in a bucket elevator boot system. Trans ASABE 55:659–672

28. González-Montellano C, Baguena EM, Ramírez-Gómez Á, Barreiro P (2014) Discrete element analysis for the assessment of the accuracy of load cell-based dynamic weighing systems in grape harvesters under different ground conditions. Comput Electron Agric 100:13–23. https://doi.org/10.1016/j.compag.2013.10.008
29. Van Zeebroeck M, Tijskens E, Dintwa E et al (2006) The discrete element method (DEM) to simulate fruit impact damage during transport and handling: model building and validation of DEM to predict bruise damage of apples. Postharvest Biol Technol 41:85–91. https://doi.org/10.1016/j.postharvbio.2006.02.007
30. Ramírez-Gómez Á, Gallego E, Fuentes JM et al (2014) Values for particle-scale properties of biomass briquettes made from agroforestry residues. Particuology 12:100–106. https://doi.org/10.1016/j.partic.2013.05.007
31. Cundall PA (1988) Detect and represent contacts in a system composed of many polyhedral blocks. Int J Rock Mech Min Sci Geomech 25:107–116
32. Hart R, Cundall PA, Lemos J (1988) Formulation of a three-dimensional distinct element model—Part II. Mechanical calculations for motion and interaction of a system composed of many polyhedral blocks. Int J Rock Mech Min Sci 25:117–125. https://doi.org/10.1016/0148-9062(88)92294-2
33. Horabik J, Molenda M (2016) Parameters and contact models for DEM simulations of agricultural granular materials: a review. Biosyst Eng 147:206–225. https://doi.org/10.1016/j.biosystemseng.2016.02.017
34. Kosinski P, Hoffmann AC (2010) An extension of the hard-sphere particle-particle collision model to study agglomeration. Chem Eng Sci 65:3231–3239. https://doi.org/10.1016/j.ces.2010.02.012
35. O'Sullivan C (2011) Particle-based discrete element modeling: geomechanics perspective. Int J Geomech 11:449–464. https://doi.org/10.1061/(asce)gm.1943-5622.0000024
36. O'Sullivan C (2011) Particulate discrete element modelling: a geomechanics perspective, 1st edn. Spon Press, New York
37. Hoomans BPB, Kuipers JAM, Briels WJ, Van Swaaij WPM (1996) Discrete particle simulation of bubble and slug formation in a two-dimensional gas-fluidized bed : a hard-sphere approach. Chem Eng Sci 51:99–118. https://doi.org/10.1016/0009-2509(95)00271-5
38. Fortin J, Millet O, de Saxcé G (2005) Numerical simulation of granular materials by an improved discrete element method. Int J Numer Methods Eng 62:639–663. https://doi.org/10.1002/nme.1209
39. Norouzi HR, Zarghami R, Sotudeh-gharebagh R, Mostoufi N (2016) Coupled CFD-DEM modeling: formulation, implementation and application to multiphase flows, 1st edn. Wiley, Newark
40. Duran J (2000) Sands, powders, and grains: an introduction to the physics of granular materials. Springer, New York
41. Tsuji Y, Kawaguchi T, Tanaka T (1993) Discrete particle simulation of two-dimensional fluidized bed. Powder Technol 77:79–87. https://doi.org/10.1016/0032-5910(93)85010-7
42. Zhao J, Shan T (2013) Coupled CFD-DEM simulation of fluid-particle interaction in geomechanics. Powder Technol 239:248–258. https://doi.org/10.1016/j.powtec.2013.02.003
43. Guo Y, Curtis JS (2015) Discrete element method simulations for complex granular flows. Annu Rev Fluid Mech 47:21–46. https://doi.org/10.1146/annurev-fluid-010814-014644
44. Campbell CS (2006) Granular material flows – an overview. Powder Technol 162:208–229. https://doi.org/10.1016/j.powtec.2005.12.008
45. Delaney GW, Inagaki S, Aste T (2007) Fine tuning DEM simulations to perform virtual experiments with three-dimensional granular packings. In: Tomaso A, Matteo T Di, Tordesillas A (eds) Granular and complex materials. World Scientific, Singapore, Hackensack, pp. 141–168
46. Carlos Varas AE, Peters EAJF, Kuipers JAM (2017) CFD-DEM simulations and experimental validation of clustering phenomena and riser hydrodynamics. Chem Eng Sci 169:246–258. https://doi.org/10.1016/j.ces.2016.08.030

47. He Y, Bayly AE, Hassanpour A (2018) Coupling CFD-DEM with dynamic meshing: a new approach for fluid-structure interaction in particle-fluid flows. Powder Technol 325:620–631. https://doi.org/10.1016/j.powtec.2017.11.045
48. Neuwirth J, Antonyuk S, Heinrich S, Jacob M (2013) CFD-DEM study and direct measurement of the granular flow in a rotor granulator. Chem Eng Sci 86:151–163. https://doi.org/10.1016/j.ces.2012.07.005
49. Akhshik S, Behzad M, Rajabi M (2015) CFD-DEM approach to investigate the effect of drill pipe rotation on cuttings transport behavior. J Pet Sci Eng 127:229–244. https://doi.org/10.1016/j.petrol.2015.01.017
50. Boac JM, Casada ME, Maghirang RG (2010) Material and interaction properties of selected grains and oilseeds for modeling discrete particles. Trans ASABE 53:1201–1216. https://doi.org/10.13031/2013.32577
51. ESSS-Rocky (2018) Rocky-Dem technical manual. Model used Rocky
52. Cundall PA (1988) Computer simulations of dense sphere assemblies. Micromech Granul Mater 20:113–123. https://doi.org/10.1016/B978-0-444-70523-5.50021-7
53. Timoshenko SP, Goodier JN (1970) Theory of elasticity, 3rd edn. McGraw-Hill, New York
54. ASAE S368.4 (2001) Compression test of food materials of convex shape, revised. ASAE, Michigan
55. Tsuji Y, Tanaka T, Ishida T (1992) Lagrangian numerical-simulation of plug flow of cohesionless particles in a horizontal pipe. Powder Technol 71:239–250. https://doi.org/10.1016/0032-5910(92)88030-L
56. Antypov D, Elliott JA (2011) On an analytical solution for the damped Hertzian spring. Europhys Lett 94:50004. https://doi.org/10.1209/0295-5075/94/50004
57. Walton OR, Braun RL (1986) Viscosity, granular-temperature, and stress calculations for shearing assemblies of inelastic, frictional disks. J Rheol (N Y N Y) 30:949–980. https://doi.org/10.1122/1.549893
58. Walton OR (1993) Numerical simulation of inelastic, frictional Particle_Particle interactions. In: Roco MC (ed) Particulate two-phase flow, 2nd edn. Adivision of Reed Publishing (USA) Inc, Wilmington, pp 855–947
59. Fonte CB Jr, JAA O, De Almeida LC (2015) Dem-Cfd coupling: mathematical modelling and case studies using rocky-Dem® and Ansys Fluent®. In: Eleventh international conference on CFD in the minerals and process industries. CSIRO, Melbourne
60. Freireich B, Litster J, Wassgren C (2009) Using the discrete element method to predict collision-scale behavior: a sensitivity analysis. Chem Eng Sci 64:3407–3416. https://doi.org/10.1016/j.ces.2009.04.019
61. Almeida E, Spogis N, Silva MA (2016) Computational study of the pneumatic separation of sugarcane bagasse. In: 6th international conference on engineering for waste and biomass valorisation
62. De Almeida E, Spogis N, Taranto OP, Silva MA (2019) Theoretical study of pneumatic separation of sugarcane bagasse particles. Biomass Bioenergy 127:105256. https://doi.org/10.1016/j.biombioe.2019.105256
63. Bartsch P, Zunft S (2019) Granular flow around the horizontal tubes of a particle heat exchanger : DEM-simulation and experimental validation. Sol Energy 182:48–56. https://doi.org/10.1016/j.solener.2019.01.086
64. Vivacqua V, López A, Hammond R et al (2019) DEM analysis of the effect of particle shape, cohesion and strain rate on powder rheometry. Powder Technol 342:653–663. https://doi.org/10.1016/j.powtec.2018.10.034
65. Seville JPK, Willett CD, Knight PC (2000) Interparticle forces in fluidisation.pdf. Powder Technol 113:261–268. https://doi.org/10.1016/S0032-5910(00)00309-0
66. Schwager T, Pöschel T (2007) Coefficient of restitution and linear-dashpot model revisited. Granul Matter 9:465–469. https://doi.org/10.1007/s10035-007-0065-z

67. Van Zeebroeck M, Tijskens E, Van Liedekerke P et al (2003) Determination of the dynamical behaviour of biological materials during impact using a pendulum device. J Sound Vib 266:465–480. https://doi.org/10.1016/S0022-460X(03)00579-0
68. Stropek Z, Gołacki K (2015) A new method for measuring impact related bruises in fruits. Postharvest Biol Technol 110:131–139. https://doi.org/10.1016/j.postharvbio.2015.07.005
69. Deresiewicz H, Mindlin RD (1953) Elastic spheres in contact under varying oblique forces. J Appl Mech 20:327–344. https://doi.org/10.1115/1.4010702
70. Vu-Quoc L, Zhang X, Walton OR (2000) A 3-D discrete-element method for dry granular flows of ellipsoidal particles. Comput Methods Appl Mech Eng 187:483–528. https://doi.org/10.1016/S0045-7825(99)00337-0
71. Di Renzo A, Di Maio FP (2004) Comparison of contact-force models for the simulation of collisions in DEM-based granular flow codes. Chem Eng Sci 59:525–541. https://doi.org/10.1016/j.ces.2003.09.037
72. Łukaszuk J, Molenda M, Horabik J, Wiącek J (2009) Method of measurement of coefficient of friction between pairs of metallic and organic objects. Acta Agrophysica 13:407–418
73. Nyembwe AM, Cromarty RD (2019) Simulation of the pressure drop across granulated mixtures using a coupled DEM—CFD model. Adv Powder Technol 30:85–97. https://doi.org/10.1016/j.apt.2018.10.010
74. Sun J, Battaglia F, Subramaniam S (2006) Dynamics and structures of segregation in a dense, vibrating granular bed. Phys Rev E 74:061307 (1–13). https://doi.org/10.1103/PhysRevE.74.061307
75. Li H, Li Y, Gao F et al (2012) CFD-DEM simulation of material motion in air-and-screen cleaning device. Comput Electron Agric 88:111–119. https://doi.org/10.1016/j.compag.2012.07.006
76. Ren B, Zhong W, Chen Y et al (2012) CFD-DEM simulation of spouting of corn-shaped particles. Particuology 10:562–572. https://doi.org/10.1016/j.partic.2012.03.011
77. Mousaviraad M, Tekeste MZ, Rosentrater KA (2017) Calibration and validation of a discrete element model of corn using grain flow simulation in a commercial screw grain auger. Trans ASABE 60:1403–1415. https://doi.org/10.13031/trans.12200
78. Han D, Zhang D, Jing H et al (2018) DEM-CFD coupling simulation and optimization of an inside-filling air-blowing maize precision seed-metering device. Comput Electron Agric 150:426–438. https://doi.org/10.1016/j.compag.2018.05.006
79. Chan EL, Washino K (2018) Coarse grain model for DEM simulation of dense and dynamic particle flow with liquid bridge forces. Chem Eng Res Des 132:1060–1069. https://doi.org/10.1016/j.cherd.2017.12.033
80. Olaleye AK, Shardt O, Walker GM, Van Den Akker H (2019) Pneumatic conveying of cohesive dairy powder: experiments and CFD-DEM simulations. Powder Technol. https://doi.org/10.1016/j.powtec.2019.09.046
81. Sakai M, Takahashi H, Pain CC et al (2012) Study on a large-scale discrete element model for fine particles in a fluidized bed. Adv Powder Technol 23:673–681. https://doi.org/10.1016/j.apt.2011.08.006
82. Sorace CM, Louge MY, Crozier MD, Law VHC (2009) High apparent adhesion energy in the breakdown of normal restitution for binary impacts of small spheres at low speed. Mech Res Commun 36:364–368. https://doi.org/10.1016/j.mechrescom.2008.10.009
83. Althaus TO, Windhab EJ, Scheuble N (2012) Effect of pendular liquid bridges on the flow behavior of wet powders. Powder Technol 217:599–606. https://doi.org/10.1016/j.powtec.2011.11.026
84. Johnson KL, Kendall K, Roberts AD (1971) Surface charge and the contact of elastic solids. Proc R Soc Lond 324:301–313. https://doi.org/10.1088/0022-3727/6/9/304
85. Nase ST, Vargas WL, Abatan AA, McCarthy JJ (2001) Discrete characterization tools for cohesive granular material. Powder Technol 116:214–223. https://doi.org/10.1016/S0032-5910(00)00398-3

Chapter 6
CFD–DEM Coupling

Abstract Employment of the coupled computational fluid dynamics and discrete element method (CFD–DEM) to the analysis of aerodynamic separation systems presents an opportunity for the deep understanding of the steps involved in the synergistic integration of CFD and DEM for the analysis and optimization of aerodynamic separation systems. This chapter describes the intricate interplay between the dynamics of the fluid flow and particulate behavior, offering an understanding of how these two methodologies converge to enhance the efficiency, accuracy, and predictive capabilities of separation technologies. By elucidating the theoretical foundations, numerical techniques, and practical applications of CFD–DEM coupling, this chapter aims to provide the essential knowledge needed to harness the full potential of this innovative computational approach in designing and optimizing the cutting-edge aerodynamic separation systems.

Keywords Granular flow · Momentum exchange · Particle-flow interaction · Simulation methods

6.1 Approaches to CFD–DEM Coupling

The novel CFD–DEM coupling methodology was first introduced by a group of pioneering researchers [1–4] and has since been widely explored and refined by numerous experts in the field. Integrating the DEM with CFD presents a highly effective and dependable technique for comprehending the intricate interplay between particles and fluid in motion, leading to the design and optimization of granular matter systems [5, 6]. In essence, granular flow refers to the movement of a group of individual solids that experience energy loss when interacting with themselves, or fluid flow [7]. Granular matter has extensive industrial applications and is frequently modeled through the CFD–DEM approach [8]. Through the CFD–DEM coupling method, CFD facilitates the generation of fluid flow data files, encompassing a diverse range of fluids such as air, water, and gases. At the same time, DEM characterizes the granular matter present within the fluid, thereby predicting its

L. Zhou et al., *Discrete Element Method for Multiphase Flows with Biogenic Particles*, https://doi.org/10.1007/978-3-031-67729-8_6

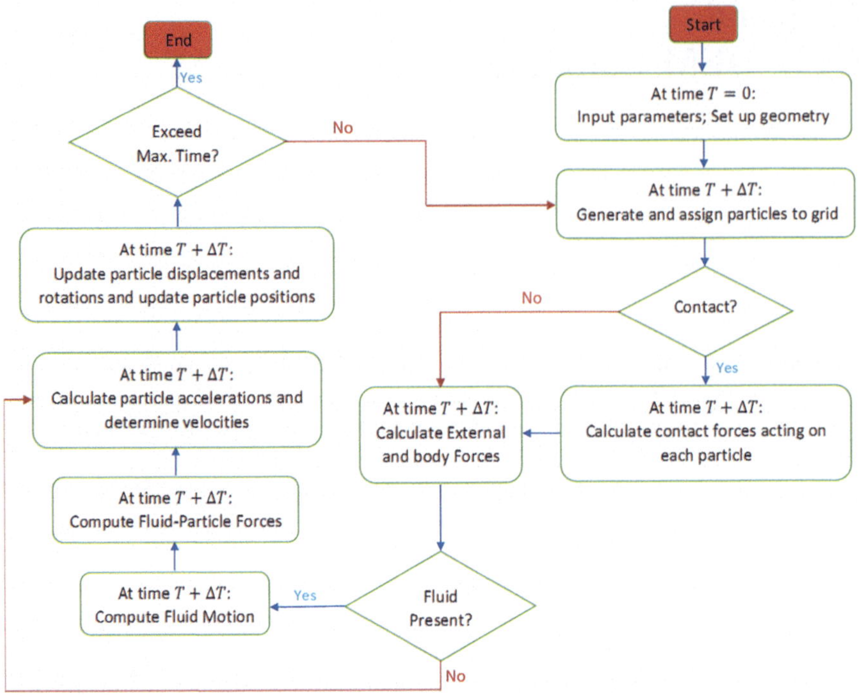

Fig. 6.1 Schematic diagram of CFD–DEM coupling technique

behavior with remarkable accuracy. Figure 6.1 illustrates the fundamental algorithmic process of the CFD–DEM coupling approach, providing a clear and concise depiction of this cutting-edge methodology.

Studying the interaction between particles and fluids has been an important area of research in many fields. In recent years, the development of the DEM and CFD has allowed for more accurate and detailed simulations of these interactions. To simulate the behavior of particles in a fluid, the DEM method assigns properties to each particle and places them into a grid cell. The CFD algorithm then resolves the fluid flow field characteristics to reach convergence. The flow's resulting velocity, pressure, density, and viscosity are then transferred to the DEM method to generate a coupling interface and compute the forces acting on each particle. This CFD–DEM coupling interface considers particles' translational and rotational motion data from the DEM solver. It computes volume fractions and momentum exchange in the mesh cell of the CFD. Depending on the turbulence effect, three main methods exist for simulating the interaction between particles and surrounding fluids [6, 9]. The first method focuses on simulating how fluids change the flow of particles with which they come into contact. The second method simulates how particles change the flow of the fluids. The last method considers the fluid streamlines compressed between the particles during the simulation. The development of various CFD–DEM approaches has dramatically improved the understanding of the interaction between particles and fluids. Interested readers can refer to the literature for a more comprehensive review of these approaches [10].

In today's world, CFD–DEM simulations have become a prevalent technique for understanding the interactions between particles and turbulence [9, 11]. The application of dimensionless coordinates to illustrate the various coupling regimes, as depicted in Fig. 6.2, has facilitated the development of numerous software packages for commercial and open-source use. These software packages include Rocky®, EDEM®, PFC™, SAMADII©, CFDEM®, LIGGGHTS®, Yade, and MFIX-DEM, all of which provide a range of flexible and versatile options for CFD-DEM coupling simulations. With Rocky® and EDEM® capable of coupling with Ansys Fluent®, PFC™ offering a built-in fluid coupling package, and SAMADII© utilizing single and multiple GPUs, these software packages cater to a diverse set of requirements. CFDEM® offers a CFD–DEM approach in C++, while LIGHHHTS and Yade provide DEM simulations in C++. MFIX-DEM is a CFD–DEM approach for fluid-solid flows in FORTRAN. With such a vast range of software packages available, researchers and industry professionals can select the best-suited package for their simulation needs, thus enabling accurate and efficient simulations.

In the flow regime diagram (Fig. 6.2), the terminology of symbols can be definedas follows: α_i is the volume fraction of particles [$\alpha_i = \frac{N \times \mathrm{Vol}_i}{\mathrm{Vol}}$, N = number ofparticles, Vol_i = particle volume, and Vol = volume occupied by particles andfluid]; T_i is the response time of particle [$T_i = \frac{\rho_i d_i}{18\rho_f \upsilon_f}$, υ_f = kinematic viscosityof the fluid]; T_K is the Kolmogorov time scale [$T_K = \sqrt{\frac{\upsilon_f}{\varepsilon}}$]; T_e is the turnovertime of a large eddy [$T_e = {}^{l}\!/_{u}$, l = length scale of the energy-containing eddies and$u$ = root mean square of fluid velocity].

6.1.1 *One-Way Coupling*

In the realm of multiphase flow, there exists a particular interval known as the flow regime where particle volume fractions are exceedingly low, reaching values of ($\alpha_i \leq 10^{-6}$). During this interval, particle effects on fluid turbulence are infinitesimal, leading to a phenomenon known as one-way coupling. Essentially, this means that particle distribution depends on the state of fluid turbulence and the momentum transfer from particles to fluid. Given the negligible effect of discrete particle concentrations, particle movement is solely influenced by the fluid flow with velocities equivalent to the surrounding fluid [9]. Numerically, the solution methodology involves CFD calculations that generate velocity and pressure distribution profiles of fluid flow throughout the domain. Since particle effects on the fluid are excluded, steady-state simulations are the only required calculations [12]. Once the flow field

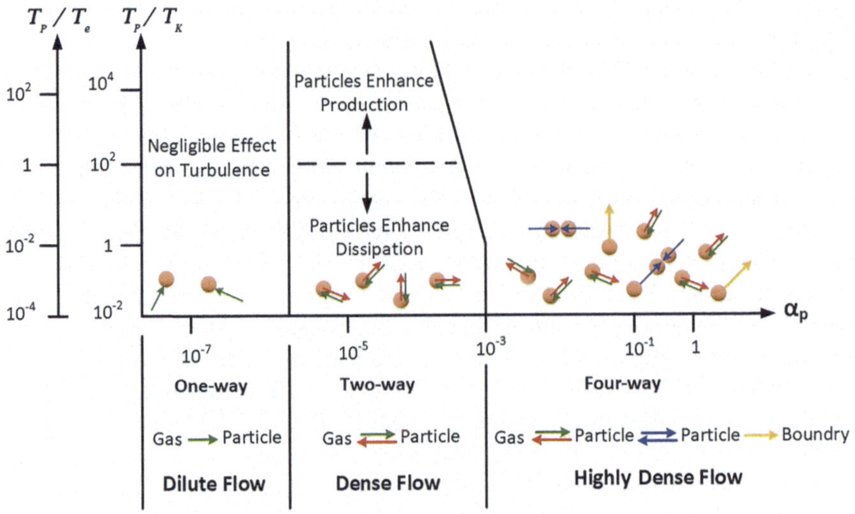

Fig. 6.2 Diagram of flow regimes depending on the interaction between particles and turbulence [9, 11]

has reached a steady state, or at the CFD simulation's conclusion, data is exported and imported into the DEM approach. By applying external forces, DEM then calculates how fluid flow would impact particle flow. This method is beneficial for large particle separation processes and unconfined homogeneous flows that vary in particle density [9, 13, 14].

In recent studies, Potapov et al. [15] utilized the moving porosity technique with DEM and CFD Ansys Fluent® using one-way coupling to simulate the multiphase flow of gas and iron particles around transfer chutes. Their technique was explicitly selected due to the large and dense nature of the particles (+4 mm). The authors' model predictions were validated against experimental results, affirming their suitability for predicting airflow around industrial-sized transfer chutes. Similarly, Fonte et al. [16] employed the one-way coupling of DEM and CFD Ansys Fluent® to separate light and heavy particles in a wind-shifter device. They used different drag laws at a fixed value of airflow and various coarse particles with geometries of paper (flat), stone (sphere), metal (rod), and wood (briquette) to determine the appropriate drag law for simulating the separation process. Their simulation showed that the Ganser drag model [17] yielded more accurate results by considering particle geometry. Finally, Almeida et al. [18, 19] utilized the one-way coupling method of DEM and CFD Ansys Fluent® to develop a theoretical algorithm for separating sugarcane bagasse into different particle groups. They varied the air velocities through an aerodynamic conveying system while considering the contact forces between particle–particle and particle–wall. The results showed that this coupling technique accurately described the separation process with reasonable accuracy. The authors recommend using coarse and large particles to increase the efficiency and realism of the separation process. Overall, these studies showcase the effectiveness of one-way coupling for simulating various multiphase flow scenarios with large and dense particles. Their findings have practical applications in industries such as mining and agriculture, where efficient separation processes are critical for optimal operations.

6.1.2 Two-Way Coupling

When the particle volume fraction falls within the range of $10^{-6} \leq \alpha_i \geq 10^{-3}$ and the particle loading volume is significant enough to impact the turbulence structure, the two-way coupling regime comes into play. This regime is characterized by the exchange of mass, momentum, and velocity information between the fluid and solid phases, resulting in a noticeable feedback effect on turbulence damping due to buoyancy effects [13]. The two-way coupling effect is a manifestation of the fact that particles are an integral part of the fluid flow, and as such, their movement is influenced not only by the fluid but also by other particles in the mix. Conversely, the flow dynamics themselves are altered by the particles, creating a feedback loop between the solid and fluid phases. Figure 6.2 illustrates how, at a specific particle volume fraction, a decrease in particle diameter, coupled with low fluid viscosity, can increase the surface area of the particulate phase, thereby increasing the dissipation rate of turbulence energy. Conversely, an increase in T_i accompanied by a constant value of α_i (inclined line) can elevate the particle Reynolds number, resulting in vortex shedding that enhances the production of turbulence energy. The two-way coupling technique utilizes a transient state simulation that enables CFD and DEM solvers to operate in parallel, significantly reducing the simulation time [12]. However, it should be noted that this method is most effective when dealing with many particles within a CFD cell, rather than particles that are larger than the CFD cells themselves. Therefore, it is advisable to use a domain mesh size larger than the largest particle size within the simulated domain [6, 20]. The primary benefit of this two-way coupling simulation is that particle agglomeration can alter airflow trajectories and velocities, and vice versa. As a result, the interaction between solids and gas is much closer to real-world phenomena [19].

There has been a growing interest in studying gas-solid flow in recent years. Researchers have been utilizing advanced simulation techniques to better understand the hydrodynamic behavior of particles in these systems. One such technique that has gained popularity is the two-way coupling approach, which considers drag, contact, and gravitational forces in three-dimensional simulations. A study by Zhong et al. [21] utilized this approach to simulate cylinder-shaped particles in a gas-solid fluidized bed and observed good agreement between predicted particle behavior and pressure drops at different superficial gas velocities with experimental results. Ren et al. [22] extended this methodology to study corn-shaped particles in a cylindrical spouted bed with a conical base device. They showed that the coupling model effectively described the hydrodynamic behavior of the spouted bed. Another study by Fonte et al. [16] applied the two-way coupling technique of DEM and CFD

Ansys Fluent® to simulate a fluidized bed and observed a good agreement between the pressure drop predicted by Ergun's correlation and the pressure drop obtained in the simulation, demonstrating the capability of the fluid being affected by a reactive force. Devarampally [13] developed a two-way coupling of CFD–DEM using the Simcenter STAR-CCM+ approach for a top spray fluidized bed granulator and successfully predicted the changes in particle velocities, temperatures, collision dynamics, and particle transfer as the inlet velocity and temperature of the air changed, providing valuable insights into the agglomeration rates in a granulation process. Furthermore, Carlos Varas et al. [23] utilized a CFD–DEM two-way model to simulate the co-current gas–particle flow in a pseudo-2D dilute flow riser and observed close agreement between the experimental data and the simulated parameters such as core-annulus flow, cluster frequency, internal solids volume fraction, and aspect ratio. Baran et al. [24] applied a fully coupled CFD-DEM two-way approach to study the effect of particle shape on fluidization in a cylindrical fluidized bed and reported consistent observations with previously published literature, providing strong evidence for applying this technique to predict the dynamic behavior of different particles in the fluidized bed [25, 26]. Lastly, Almeida et al. [19] simulated dry bagasse separation in a pneumatic device using DEM and CFD Ansys Fluent® with a two-way coupling approach and observed high accuracy in predicting the particle separation phenomena compared to actual experiments. Overall, the application of the two-way coupling technique in gas–solid fluidized bed simulations has proven to be a reasonable and practical approach for predicting the hydrodynamic behavior of particles in different systems, providing valuable insights into the dynamics of these systems and contributing to the development of hybrid CFD–DEM–PBM models that describe the rate processes in granulation and understand the effect of process parameters on the product quality attributes.

6.1.3 Four-Way Coupling

The fascinating interplay between particles and turbulence in a fluid medium gives rise to the complex four-way coupling phenomenon. This arises from the intricate interactions between particle–fluid and particle–particle and the resulting frictions [9]. Under high particle loading and concentration, where the particle volume fraction exceeds 10^{-3}, fluid streamlines between particles become compressed, creating a dense flow regime. Despite the challenging nature of this regime, most studies have traditionally focused on dilute flows, neglecting their full complexities. Interestingly, the line separating two-way and four-way couplings appears inclined in Fig. 6.2, reflecting the collision of particles at higher values of T_i/T_e. This collision leads to a transformation from two-way to four-way coupling techniques. Moreover, as the particle volume fraction approaches unity, a total granular flow manifests where no fluid is found between particles. A parallel implementation of CFD and DEM simulates this coupling approach, effectively stimulating the two-phase fluid–solid phase. This powerful approach can be applied to numerous scenarios, unlocking insights into the complex interactions between particles and fluids.

Laín and Sommerfeld [27] utilized the Euler-Lagrange approach, accounting for two-way and four-way coupling, to predict the dynamics of the gas–particle flow. Through careful consideration of particle–wall interactions and inter-particle collisions, they found that their numerical simulations could accurately replicate observed trends. However, they caution that the neglect of wall roughness effects in numerical predictions of wall-bounded flows can lead to improper application. Similarly, Kuang et al. [28] presented a 3D numerical study of vertical aerodynamic conveying using a four-way CFD-DEM approach for dilute and dense gas–particle flow regimes. They observed that their simulation technique satisfactorily captured particle flow patterns and gas pressure characteristics. Moreover, they found that particle–particle and particle–wall forces increase with increasing particle–fluid force in dense-phase flow but remain small or negligible in dilute-phase flow.

To expand upon these findings, Sturm et al. [29] extended their previous work to demonstrate the applicability of their simulation model to different flow regimes, including dense vertical aerodynamic conveying and dilute conveying in a horizontal-to-vertical pipe bend. Furthermore, employing the Euler-Lagrange approach and an in-house soft sphere DEM technique, Traoré et al. [30] developed a complete four-way coupling model of a gas–solid 3D fluidized bed by accounting for particle–particle, fluid–particle, and particle–wall interactions. Their simulation considered a highly dense flow of 2.7 million particles in two regimes, namely, pulsed jet and bubbly. Their results revealed the existence of complex structures, such as the worm-like shape, which was like those previously described in the literature. These findings provide strong evidence supporting the use of this coupling scheme to predict sensitive phenomena in highly dense granular systems.

6.2 Drag Correlation Models

During the application of CFD–DEM coupling to an aerodynamic system, drag correlation models are of vital importance besides pressure gradient forces, especially for describing and dealing with the interaction between particle shape and concentration and the surrounding fluid [31–33]. For successful coupling, it is important to choose a suitable drag correlation model [16]. There are several drag correlation models that have been developed and are available in the literature. Most of them depend mainly on the Reynolds number, while others need other properties, such as fluid volume fraction and particle sphericity. Here, attention is concentrated on the most common correlations that can be used in modeling agricultural grains based on flow regimes. The attempts by investigators and the developments for models listed in Table 6.1 illustrate the applicability of this approach.

Table 6.1 Drag correlation models used in CFD–DEM coupling

Flow regime	Models and reference	Equations	Notes
Dilute flow	Schiller and Naumann (1933) [16, 34–38]	$C_D = \begin{cases} \frac{24}{R_{ei}}\left(1+0.15R_{ei}^{0.687}\right) & R_{ei} \le 1 \\ \max\left[\frac{24}{R_{ei}}\left(1+0.15R_{ei}^{0.687}\right)\right] & 1 < R_{ei} \le 1000 \\ 0.44 & R_{ei} > 1 \end{cases}$	Spherical particles
	Dallavalle (1948) [39]	$C_D = \left(0.63 + \frac{4.8}{\sqrt{R_{ei}}}\right) R_{ei} < 3000$	Spherical particles
	Haider and Levenspiel (1989) [40]	$C_D = \frac{24}{R_{ei}}\left(1 + A R_{ei}^{B}\right) + \frac{C}{1+\frac{D}{R_{ei}}} R_{ei} < 2.6\times10^5$ For spherical particles: $A = 0.1806$, $B = 0.6459$, $C = 0.4251$, $D = 6880.95$ For non-spherical particles: $A = \exp\left(2.3288 - 64581\varnothing_i + 2.4486\varnothing_i^2\right)$ $B = 0.0964 + 0.5565\varnothing$ $C = \exp\left(4.905 - 13.8944\varnothing_i + 18.4222\varnothing_i^2 - 10.2599\varnothing_i^3\right)$ $D = \exp\left(1.4681 + 12.2584\varnothing_i - 20.7322\varnothing_i^2 + 15.8855\varnothing_i^3\right)$	Isometric non-spherical particles and non-isometric non-spherical particles

	Ganser (1993) [16–19, 41]	$\frac{C_d}{K_2} = \frac{24}{R_{ei}K_1K_2}\left[1+0.1118\left(R_{ei}K_1K_2\right)^{0.6567}\right] + \frac{0.4305}{1+\frac{3305}{R_{ei}K_1K_2}}$ $K_1 = \left(\frac{d_i}{3d_{i_{vol}}} + \frac{2}{3}\varnothing_i^{\frac{-1}{2}}\right)^{-1} - 2.25\frac{d_{i_{vol}}}{D}$ $K_2 = 10^{1.8148\left(-\log_{10}\varnothing_i\right)^{0.5743}}$	Valid when $R_{ei}K_1K_2 < 10^5$, used when the particles in the domain differ in shape, concentration, and alignment
Dense flow	Wen and Yu (1966) [42, 43]	$C_D = \begin{cases} \frac{24}{\alpha_f R_{ei}}\left[1+0.15\left(\alpha_f R_{ei}^{0.687}\right)^{0.687}\right] \alpha_f^{-1.65} & \alpha_f R_{ei} < 1000 \\ 0.44\alpha_f^{-1.65} & \alpha_f R_{ei} \geq 1000 \end{cases}$	Relatively low particle concentration: $\alpha_i < 0.2$
	Ergun (1958) [31, 35, 44, 45]	$C_D = 200\frac{\alpha_p}{\alpha_f \varnothing_i^2 R_{ei}} + \frac{7}{3\varnothing_i}$	High particle concentration: $\alpha_i \geq 0.2$.
	Huilin and Gidaspow (2003) [46]	$C_D = \varphi C_{D,\text{ Ergun}} + (1-\varphi)C_{D,\text{ Wen \& Yu}}$ $\varphi = \frac{\arctan\left[262.5\left(0.8-\alpha_f\right)\right]}{\pi} + 0.5$	Wide range of particle concentration (low and high particle concentration): $0.2 > \alpha_i \geq 0.2$
	Di Felice (1994) [12, 19, 35, 45, 47]	$C_D = \alpha_f^{2-\beta} C_{D,\text{single particle}}$ $\beta = 3.7 - 0.65\exp\left\{-\frac{\left[1.5-\log_{10}\left(\alpha_f R_{ei}\right)\right]^2}{2}\right\}$	$10^{-2} \leq R_{ei} \leq 10^4$

6.3 Summary

Coupling CFD and DEM has revolutionized our ability to study the intricate interactions between particles and fluids. This innovative approach, pioneered by dedicated researchers and continually refined by experts, offers a powerful tool for understanding granular flow phenomena and optimizing various systems involving particulate matter. The one-way coupling regime, characterized by low particle volume fractions, allows for simulations where particle effects on fluid turbulence are minimal. This method has successfully studied extensive particle separation processes and unconfined homogeneous flows, as demonstrated by studies focusing on industrial applications such as transfer chutes and particle separation devices. Moving into the two-way coupling regime, where particle volume fractions increase, we encounter a more complex interaction between particles and the fluid. This regime allows for the exchange of mass, momentum, and velocity information between the fluid and solid phases, leading to a feedback loop that significantly influences turbulence damping and particle behavior. Applications of this method have yielded valuable insights into gas–solid fluidized beds, granulation processes, and particle dynamics in various systems. The pinnacle of complexity lies in the four-way coupling regime, where the interplay between particles and turbulence reaches a highly intricate level. Here, we observe a total granular flow where no fluid exists between particles, leading to a comprehensive understanding of dense flow regimes. Studies in this area have shed light on phenomena such as particle–wall interactions, inter-particle collisions, and complex flow structures in granular systems. Throughout these regimes, the choice of drag correlation models plays a crucial role in accurately describing the interaction between particle shape, concentration, and the surrounding fluid. Researchers and industry professionals can access various software packages and models, enabling them to select the most suitable approach for their simulation needs.

In essence, the CFD–DEM coupling methodology stands as a cornerstone in studying particle–fluid interactions, offering a robust framework for simulating, and understanding diverse systems. From industrial processes to environmental applications, this approach continues to drive innovation, improve efficiency, and enhance our comprehension of granular matter dynamics. As we look to the future, further advancements in CFD–DEM coupling promises even deeper insights into the behavior of particles in fluid environments, paving the way for continued progress in various fields.

References

1. Tsuji Y, Kawaguchi T, Tanaka T (1993) Discrete particle simulation of two-dimensional fluidized bed. Powder Technol 77:79–87. https://doi.org/10.1016/0032-5910(93)85010-7
2. Tsuji Y, Tanaka T, Ishida T (1992) Lagrangian numerical-simulation of plug flow of cohesionless particles in a horizontal pipe. Powder Technol 71:239–250. https://doi.org/10.1016/0032-5910(92)88030-L

3. Hoomans BPB, Kuipers JAM, Briels WJ, Van Swaaij WPM (1996) Discrete particle simulation of bubble and slug formation in a two-dimensional gas-fluidized bed: a hard-sphere approach. Chem Eng Sci 51:99–118. https://doi.org/10.1016/0009-2509(95)00271-5
4. Xu BH, Yu AB (1997) Numerical simulation of the gas–solid flow in a fluidized bed by combining discrete particle method with computational fluid dynamics. Chem Eng Sci 52:2785–2809. https://doi.org/10.1016/S0009-2509(97)00081-X
5. Zhao J, Shan T (2013) Coupled CFD-DEM simulation of fluid-particle interaction in geomechanics. Powder Technol 239:248–258. https://doi.org/10.1016/j.powtec.2013.02.003
6. Norouzi HR, Zarghami R, Sotudeh-gharebagh R, Mostoufi N (2016) Coupled CFD-DEM modeling: formulation, implementation and application to multiphase flows, 1st edn. Wiley, Newark
7. Duran J (2000) Sands, powders, and grains: an introduction to the physics of granular materials. Springer, New York
8. Wang Z, Teng Y, Liu M (2019) A semi-resolved CFD–DEM approach for particulate flows with kernel based approximation and Hilbert curve based searching strategy. J Comput Phys 384:151–169. https://doi.org/10.1016/j.jcp.2019.01.017
9. Elghobashi S (1994) On predicting particle-laden turbulent flows. Appl Sci Res 52:309–329. https://doi.org/10.1007/BF00936835
10. Zhu HP, Zhou ZY, Yang RY, Yu AB (2007) Discrete particle simulation of particulate systems: theoretical developments. Chem Eng Sci 62:3378–3396. https://doi.org/10.1016/j.ces.2006.12.089
11. Elghobashi S (1991) Particle-laden turbulent flows: direct simulation and closure models. Appl Sci Res 48:301–314. https://doi.org/10.1007/BF02008202
12. Zhao H, Zhao Y (2019) CFD–DEM simulation of pneumatic conveying in a horizontal channel. Int J Multiph Flow 118:64–74. https://doi.org/10.1016/j.ijmultiphaseflow.2019.06.003
13. Devarampally DR (2017) Development of two-way coupled CFD—DEM model for top spray fluid bed granular using Star CCM+. The State University of New Jersey, New Brunswick
14. Elsayed K, Lacor C (2014) Analysis and optimisation of cyclone separators geometry using RANS and LES methodologies. Turbul Interact Notes Numer Fluid Mech Multidiscip Des 125:65–74. https://doi.org/10.1007/978-3-662-43489-5
15. Potapov A, Chen X, Donohue T, Wheeler C (2013) Computer simulation of airflow around transfer chutes via linked discrete element method – computational fluid dynamics approach. In: 11th international congress on bulk materials storage, handling and transportation
16. Fonte CB Jr, JAA O, De Almeida LC (2015) Dem-Cfd coupling: mathematical modelling and case studies using Rocky-Dem® and Ansys fluent®. In: Eleventh international conference on CFD in the minerals and process industries. CSIRO, Melbourne
17. Ganser GH (1993) A rational approach to drag prediction nonspherical particles. Powder Technol 77:143–152. https://doi.org/10.1016/0032-59109(93)80051-b
18. Almeida E, Spogis N, Silva MA (2016) Computational study of the pneumatic separation of sugarcane bagasse. In: 6th international conference on engineering for waste and biomass valorisation
19. De Almeida E, Spogis N, Taranto OP, Silva MA (2019) Theoretical study of pneumatic separation of sugarcane bagasse particles. Biomass Bioenergy 127:105256. https://doi.org/10.1016/j.biombioe.2019.105256
20. ESSS-Rocky (2018) Rocky-Dem Technical Manual. In: Models used in Rocky. ESSS, Rio de Janeiro, Brazil
21. Zhong WQ, Zhang Y, Jin B, Zhang M (2009) Discrete element method simulation of cylinder-shaped particle flow in a gas-solid fluidized bed. Chem Eng Technol 32:386–391. https://doi.org/10.1002/ceat.200800516
22. Ren B, Zhong W, Chen Y et al (2012) CFD-DEM simulation of spouting of corn-shaped particles. Particuology 10:562–572. https://doi.org/10.1016/j.partic.2012.03.011
23. Carlos Varas AE, Peters EAJF, Kuipers JAM (2017) CFD-DEM simulations and experimental validation of clustering phenomena and riser hydrodynamics. Chem Eng Sci 169:246–258. https://doi.org/10.1016/j.ces.2016.08.030
24. Baran O, Eppinger T, Han K (2018) DEM simulation of cylinders and capsules in a fluidized bed. In: 8th world congress on particle technology, Orlando, pp 232–242

25. Mahajan V V., Nijssen TMJ, Fitzgerald BW, et al (2017) Fluidization of spherocylindrical particles. In: 8th international conference on micromechanics on granular media
26. Buist KA, Jayaprakash P, Kuipers JAM et al (2017) Magnetic particle tracking for nonspherical particles in a cylindrical fluidized bed. AICHE J 63:5335–5342. https://doi.org/10.1002/aic.15854
27. Laín S, Sommerfeld M (2008) Euler/Lagrange computations of pneumatic conveying in a horizontal channel with different wall roughness. Powder Technol 184:76–88. https://doi.org/10.1016/j.powtec.2007.08.013
28. Kuang SB, Yu AB, Zou ZS (2009) Computational study of flow regimes in vertical pneumatic conveying. Ind Eng Chem Res 48:6846–6858. https://doi.org/10.1063/1.3179812
29. Sturm M, Wirtz S, Scherer V, Denecke J (2010) Coupled DEM-CFD simulation of pneumatically conveyed granular media. Chem Eng Technol 33:1184–1192. https://doi.org/10.1002/ceat.201000162
30. Traoré P, Laurentie JC, Dascalescu L (2015) An efficient 4 way coupling CFD-DEM model for dense gas-solid particulate flows simulations. Comput Fluids 113:65–76. https://doi.org/10.1016/j.compfluid.2014.07.017
31. Zhong W, Yu A, Liu X et al (2016) DEM/CFD-DEM modelling of non-spherical particulate systems: theoretical developments and applications. Powder Technol 302:108–152. https://doi.org/10.1016/j.powtec.2016.07.010
32. Carruthers AC, Filippone A (2005) Aerodynamic drag of streamers and flags. J Aircr 42:976–982. https://doi.org/10.2514/1.9754
33. Gent SP, Gent SP (2009) Computational modeling of multiphase fibrous flows for simulation based engineering. Iowa State University, Ames
34. Schiller L, Naumann AZ (1933) Ueber die grundlegenden Berechnungen bei der Schwerkraftaufbereitung. Ver Deut Ing 77:317–320
35. Crowe CT, Schwarzkopf JD, Sommerfeld M, Tsuji Y (2011) Multiphase flows with droplets and particles, 2th edn. CRC Press, Taylor & Francis Group, Boca Raton
36. Pritchard PJ, Leylegian JC (2011) Fox and McDonald's introduction to fluid mechanics, 8th edn. Wiley, New York
37. Parvaz F, Hosseini SH, Elsayed K, Ahmadi G (2020) Influence of the dipleg shape on the performance of gas cyclones. Sep Purif Technol 233:116000. https://doi.org/10.1016/j.seppur.2019.116000
38. Liu R, Zhou Z, Xiao R, Yu A (2019) CFD-DEM modelling of mixing of granular materials in multiple jets fluidized beds. Powder Technol. https://doi.org/10.1016/j.powtec.2019.08.002
39. Dallavalle JM (1948) Micromeritics: the technology of fine particles, 2nd edn. Pitman Publishing Corp, New York
40. Levenspiel O, Haider A (1989) Drag coefficient and terminal velocity of spherical and nonspherical particles. Powder Technol 58:63–70. https://doi.org/10.1016/0032-5910(89)80008-7
41. ESSS-Rocky (2016) Particle drag laws implemented in the Rocky DEMCFD oneway and twoway coupling modules. In: Rocky E (ed) Rocky-Dem technical manual, 4.13. Rocky DEM, Inc, Rio de Janeiro, pp 1–9
42. Wen CY, Yu YH (1966) Mechanics of fluidization. Chem Eng Prog Symp Ser 162:100–111
43. Benzarti S, Mhiri H, Bournot H (2012) Drag models for simulation gas-solid flow in the bubbling fluidized bed of FCC particles. Int J Chem Mol Eng 6:111–116
44. Ergun S, Orning AA (1949) Fluid flow through packed columns and fluidized bed. Ind Eng Chem 41:1179–1184. https://doi.org/10.1021/ie50474a011
45. Knight C, O'Sullivan C, van Wachem B, Dini D (2020) Computing drag and interactions between fluid and polydisperse particles in saturated granular materials. Comput Geotech 117:103210. https://doi.org/10.1016/j.compgeo.2019.103210
46. Huilin L, Gidaspow D (2003) Hydrodynamics of binary fluidization in a riser: CFD simulation using two granular temperatures. Chem Eng Sci 58:3777–3792. https://doi.org/10.1016/S0009-2509(03)00238-0
47. Di Felice R (1994) The voidage function for fluid-particle interaction systems. Int J Multiph Flow 20:153–159. https://doi.org/10.1016/0301-9322(94)90011-6

Chapter 7
CFD–DEM Applications

Abstract The focus of this chapter is on application of the computational fluid dynamics with discrete element method (CFD–DEM) in simulating the complex multiphase flow field of cyclone separators for augmenting their efficiency and efficacy. The cyclone separators are used as a cornerstone equipment in various industries for particle separation. By examining the principles, methodologies, and recent advancements in CFD-DEM, an understanding of the complex flow dynamics and particle behavior inside the cyclones is investigated. Through exploration of several case studies, simulations, and experimental validations, this chapter elucidates the transformative potential of CFD–DEM in optimizing the cyclone design, enhancing its separation efficiencies, and minimizing the energy consumption. From elucidating the intricate interplay of forces governing particle separation to providing insights into the design intricacies for diverse applications, this chapter serves as a comprehensive guide to demonstrate the full potential of CFD–DEM in the realm of cyclone separator optimization.

Keywords Cyclone separator · Optimization · Particle separation · Cyclone performance

7.1 Introduction

Computational fluid dynamics with discrete element method (CFD-DEM) is a potent numerical technique that synergistically combines the inherent advantages of two distinct methodologies to simulate intricate multiphase flows. While CFD primarily focuses on modeling fluid dynamics, DEM is specifically tailored to simulate the behavior of discrete particles. The integration of these methods results in a comprehensive tool that offers profound insights into systems involving both fluid and solid phases. This approach finds widespread applications across diverse industries. A prominent application of CFD–DEM lies in the domain of granular flows, where the interplay between solid particles and fluid plays a pivotal role. This is

L. Zhou et al., *Discrete Element Method for Multiphase Flows with Biogenic Particles*, https://doi.org/10.1007/978-3-031-67729-8_7

particularly pertinent in the pharmaceuticals, food processing, agriculture, and powder metallurgy industries. A nuanced understanding of the dynamics governing granular materials during processing is indispensable for optimizing manufacturing processes, ensuring product quality, and minimizing equipment wear.

Mixing and blending powders in pharmaceutical manufacturing is a pivotal stage in drug formulation [1, 2]. Employing CFD–DEM simulations proves instrumental in unraveling intricate details of flow patterns, segregation phenomena, and powder distribution within mixing equipment. Such insights are invaluable for the meticulous design of efficient mixing processes, ensuring a uniform drug content in the final product [3–5]. The utility of CFD–DEM transcends industrial applications, extending into environmental studies, specifically in comprehending sediment transport and riverbed erosion [6, 7]. Through simulations that replicate the interplay between water flow and sediment particles, researchers gain a deeper understanding of river morphodynamics, enabling predictions of changes in riverbed profiles over time. The ramifications of this extend to effective water resource management, erosion prevention, and the preservation of ecological equilibrium in aquatic ecosystems [8–10]. In the oil and gas or petroleum industry, CFD–DEM is crucial in simulating the flow of fluids laden with particles within pipelines [11–13]. A comprehensive grasp of solid particle behavior in oil and gas transportation is indispensable for optimizing pipeline design, preventing clogging issues, and ensuring the seamless transport of hydrocarbons [14]. Furthermore, CFD–DEM applications extend to the study of particle settling in oil storage tanks, aiding in the design of efficient separation systems [15]. The scope of CFD–DEM is not confined to industrial and environmental domains but permeates into the biomedical field. Within biomedical engineering, CFD–DEM simulations effectively model blood flow and the transport of particulate matter within the human circulatory system [16, 17]. This facilitates a detailed examination of the impact of medical devices, such as stents or artificial heart valves, on blood flow patterns and the potential for particle-induced damage [18, 19]. Additionally, CFD–DEM is indispensable in simulating multiphase flows within nuclear engineering. Understanding the behavior of particulate matter in nuclear reactors is paramount to ensuring the safety and efficiency of these facilities [20]. Through CFD–DEM simulations, the transport of fuel particles, coolant flow, and heat transfer within nuclear reactors is modeled, contributing significantly to the design and optimization of nuclear power systems [21, 22].

Despite numerous advantages and diverse applications, CFD–DEM has emerged as a potent tool in agricultural research, offering a sophisticated approach to understanding and optimizing various processes within the agricultural domain [23]. A notable application of CFD–DEM in agriculture is the analysis of mechanisms for seed and fertilizer distribution [24]. By simulating the fluid dynamics involved in seed and fertilizer dispersion, researchers can gain insights into the efficacy of different seeding and fertilization techniques, optimizing resource utilization and enhancing crop yield [25–28]. In precision agriculture, where efficient resource management is paramount, CFD–DEM plays a crucial role in assessing the performance of various agricultural machinery. For instance, simulating airflow around agricultural machineries, such as tractors, combine harvesters, plowing machines, and planters, enables researchers to evaluate aerodynamic efficiency and identify areas for improvement [29–33]. This knowledge contributes to designing more aerodynamic, streamlined equipment, reducing energy consumption, and improving overall operational efficiency. Moreover, CFD–DEM aids in studying soil-structure interactions, providing valuable information on the impact of tillage practices and soil cultivation on crop growth [34–38]. Understanding the dynamic interplay between soil particles and agricultural tools helps optimize tillage strategies, minimize soil erosion, and enhance soil health. This holistic approach to agricultural research underscores the importance of CFD–DEM in advancing the understanding of processes critical to sustainable and efficient agricultural practices.

The application of CFD–DEM in agriculture extends significantly to modeling grain drying processes [39–41]. This involves simulating fluid dynamics and heat transfer to optimize drying facilities and protocols, ensuring efficient moisture removal while preserving grain quality. This is particularly crucial in post-harvest handling, where effective drying is essential to prevent mold growth and maintain the nutritional integrity of harvested crops. Through CFD–DEM simulations, researchers can explore diverse drying techniques and configurations, fostering the development of energy-efficient and cost-effective grain drying systems.

In the pesticide and herbicide application realm, CFD–DEM provides valuable insights into the dispersion patterns of sprayed particles [42, 43]. A comprehensive understanding of pesticide droplet behavior in various environmental conditions allows for refining spraying techniques, ensuring optimal coverage and effectiveness. This refinement, in turn, has implications for crop protection and environmental sustainability, as precise application methods reduce the quantity of agrochemicals required, thereby minimizing the environmental impact. Furthermore, CFD–DEM can be applied to study interactions between sprayed particles and crop canopies, facilitating the development of improved strategies for targeted pesticide delivery [44–46]. The integration of CFD–DEM in irrigation system design has revolutionized water distribution in agriculture. By simulating water flow through irrigation systems and its interaction with soil and crops, researchers can optimize irrigation practices to achieve uniform water distribution and minimize wastage [47–50]. This contributes to water conservation efforts and promotes sustainable agricultural practices. The capability to model sediment transport in irrigation water additionally plays a crucial role in preventing soil erosion, thereby sustaining the long-term productivity of agricultural land. Moreover, the integration of CFD–DEM plays a crucial role in examining the dynamics of agricultural particles, including granular materials and powders. This is particularly pertinent in designing and optimizing handling and processing systems for bulk materials in the agricultural industry [51–55]. Whether addressing the flow of grains through silos or the mixing of powdered fertilizers and packing, CFD–DEM simulations offer invaluable insights into particle behavior [56–58]. This capability empowers engineers to conceptualize equipment that maximizes efficiency and mitigates the risk of clogging or segregation.

In conclusion, the diverse and impactful applications of CFD–DEM in agriculture span various stages of the agricultural process. Integrating CFD–DEM enriches our comprehension of intricate agricultural systems. This spans precision farming and machinery design to post-harvest handling and irrigation systems. Such a multidisciplinary approach contributes significantly to heightened productivity and resource efficiency while aligning with the evolving demands of modern agriculture. As technology progresses, the role of CFD–DEM in agriculture is poised to expand further, offering innovative solutions to address the intricate challenges confronted by the global agricultural community.

7.2 Cyclone Separator Application

7.2.1 Descriptions of Simulation and Flow Conditions

Efficient separation of air and biogenic particles is crucial in many industrial applications. Reverse-flow cyclone separators are popular due to their simplicity and effectiveness. However, to optimize their performance, accurate simulations are necessary. The application reported here used a one-way coupling approach of CFD–DEM to simulate the multiphase flow in reverse-flow cyclone separators. The simulations were conducted on a high-performance desktop with an Intel(R) Xenon(R) CPU E5-2620 v4 @ 2.10 GHz and 64 GB of RAM. Three simulation cases were performed to validate and optimize the reverse flow cyclone separators with true-shape modeling of particulate matter flow. In the first case, the performance of a designed aerocyclone with different particle mass loading was simulated and validated based on the experimental work by El-Emam et al. [59]. In the second case, the designed aerocyclone separator performance was optimized by modulating and modifying the geometry dimensions. Finally, the performance of standard aerocyclones (namely, 1D3D, 2D2D, 1D2D, and 1D3D, with 2D2D inlet) was predicted and optimized, and a performance comparison was conducted between the optimized standard aerocyclones and the optimized aerocyclone separator.

7.2.2 Construction of Geometrical Models

Cyclones are well-known weather phenomena, and their geometries can now be accurately recreated using the cutting-edge UG Graphics Nx 10 CAD software. These models can then be meshed using the ANSYS ICEM CFD 18.0 software, allowing for detailed analysis of the fluid flow parameters. The result files are then saved in a (.mesh) format and are seamlessly exported to CFD Fluent® software, where the simulation process begins. Interestingly, research has shown that the computational domain of the investigated cyclones can be divided arbitrarily by unstructured tetrahedra CFD grid cells [60, 61]. This grid type does not significantly affect the simulation results, especially when the flow is regarded as turbulent. It turns out that one can rely on the accuracy of these simulations to gain valuable insights into the behavior of these powerful natural phenomena.

7.2.2.1 Designed Cyclone

The designed cyclone is composed of a volute inlet, which facilitates the introduction of air-particle flow, and a barrel section that culminates in a conical component. A seed outlet is fixed to the bottom of the cone, while the top barrel contains an outlet tube to discharge the separated leaves. The cyclone design presented here is informed by the previous literature [62–66]. The cyclone model was designed with dimensions referenced in Table 7.1 to achieve the intended geometry. The model was arbitrarily meshed with high-resolution three-dimensional tetrahedral meshes totaling 367,198 grid cells. The schematic 3D geometry and CFD mesh cells used in the simulation with DEM boundaries are presented in Figure 7.1a, b, respectively.

7.2.2.2 Optimized Cyclones

Four different cyclone geometries (II, III, IV, V, VI) were successfully created and modified compared to the designed one (I) to optimize the performance of the designed cyclone separator. The modifications to the designed cyclone (I) were done by the inclusion of an air inlet portion on the barrel (II), air inlet dimensions (III, IV), and barrel length (V). For simulations, three-dimensional tetrahedral unstructured meshes with high resolution were generated for the optimized cyclones using the ICEM CFD software as depicted in Fig. 7.2. The total number of mesh cells and the dimensions of the five optimized cyclones are shown in Tables 7.2 and 7.3.

7.2.2.3 Standard Cyclones

Four standard geometries of the reverse flow cyclone separators, namely, 1D2D, 2D2D, 1D3D, and 1D3Dw/2D2D inlet, were modeled and employed in this simulation case to compare the performance with the optimized cyclones. The cyclones had a boundary wall, flow inlet section, product outlet, and vortex finder. They were designed with the same diameter of 320 mm as the designed one. The four geometries were successfully created in the Nx-CAD software and saved as a (.x_t) formatted file. The dimensions of the cyclones, geometries, and the computational mesh and numbers are illustrated in Figs. 7.3 and 7.4 and Tables 7.4 and 7.5.

Table 7.1 The dimensions and the components of the designed cyclone separator

Component	Barrel diameter	Barrel length	Cone length	Vortex diameter	Inlet diameter	Inlet height	Inlet width	Vortex length	Inlet position from the top	Outlet tip
Geometry	D_b	L_b	L_c	D_v	D_i	H	W	L_v	Z	D_o
Dimension, mm	320	420	750	120	100	125	95	250	70	100

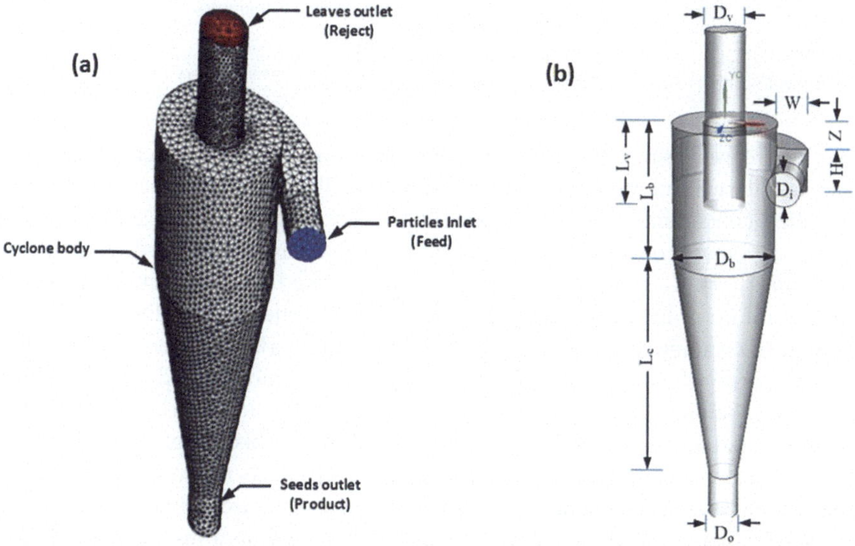

Fig. 7.1 The designed reverse-flow cyclone separator: (**a**) CFD grid with DEM components and (**b**) cyclone dimensions

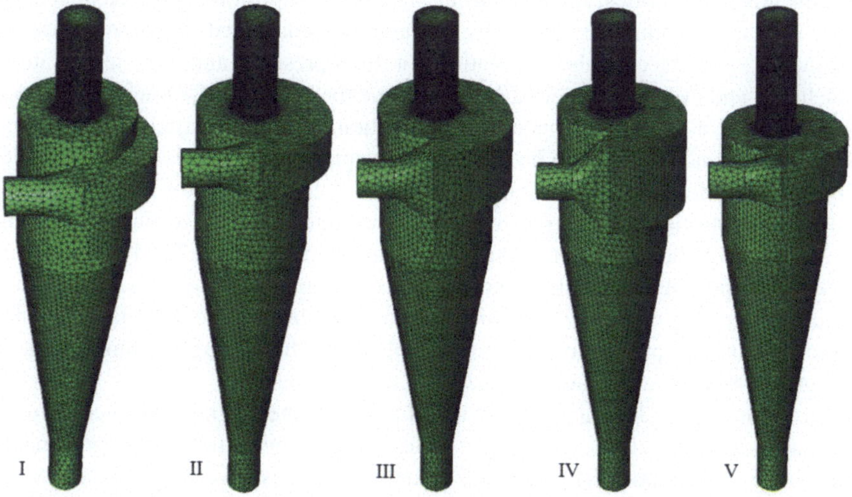

Fig. 7.2 Meshed domains of the optimized cyclones including the designed one

7.2.3 Particles Modeling

Crops are the primary source of various biogenic particulate matter, including cereal grains and seeds, which exhibit a wide range of physical characteristics such as shape, size, and density. These seeds and grains, considered the focal point of this

Table 7.2 The total mesh cells number in the optimized cyclone separators

Cyclone type	I	II	III	IV	V
Total grid number	367,198	366,658	372,973	377,221	346,086

inquiry, are used as raw materials for milling and food, providing essential nutrients such as calories and protein in human and animal diets. However, these vital crops are often contaminated by heterogeneous masses such as stones, sticks, weeds, chaff, and other crop seeds, which are introduced during the harvesting and post-harvesting [67]. Recently, the Jojoba plant has garnered global attention as a commercial oil crop due to its seeds' high liquid wax content, exceeding 50% of its weight, often referred to as an "oil" that bears a remarkable resemblance to the famed sperm whale oil [68]. The Jojoba plant (Simmondsia Chinensis (Link) Schneider) belongs to the Simmondsiaceae family, native to the arid and semi-arid zones of Arizona, United States, and Northern Mexico. This evergreen woody desert shrub grows slowly but can live up to 150 years [69]. As the world's population continues to grow exponentially, the importance of agriculture cannot be overstated. The Jojoba plant's potential as a commercial oil crop presents a sustainable and eco-friendly solution to meet the world's growing demand for essential oils.

The Jojoba plant, with its distinct pronunciation of (Ho-Ho-BA), has emerged as a valuable renewable source of high-quality oil. Unlike other vegetable oils, predominantly triglycerides, Jojoba oil offers a unique composition with potential benefits for various industries. Extensive research has confirmed its potential use in cosmetics, pharmaceuticals, lubrication for high-pressure and high-temperature machines, and even as an alternative fuel source that may surpass traditional diesel fuel [70]. In a recent computational simulation, biogenic particles comprising Jojoba seeds and leaves were investigated, as depicted in Fig. 7.5. As the seeds do not mature uniformly, they naturally fall onto the soil surface at different times and mix with impurities, including leaves and stones. Pneumatic harvesting, in particular, poses a significant challenge due to these impurities [59, 71, 72]. However, this simulation did not account for sand or stones of similar size to the seeds, as this would have resulted in increased computational simulation time. In practical applications, a sieving process can be implemented after harvesting to eliminate such impurities, thereby eliminating the need for their simulation.

In summary, the Jojoba plant and its oil have attracted significant interest from various industries, with several studies attesting to its potential. While pneumatic harvesting poses some challenges due to impurities, it is reassuring that a sieving process can efficiently address this issue. As such, Jojoba oil remains a promising renewable resource that is worth exploring further.

Jojoba seeds, depicted in Fig. 7.5b, possess a unique and striking appearance. These small, hard, dark brown seeds are elongated with a hint of sphericity and have a wrinkled surface resembling a coffee bean. The seeds are polymorphic, varying in size, shape, color, and weight, depending on the form of the ovary, the size of the embryo, the environmental conditions during seed formation, and the species itself [73]. Interestingly, seed size is affected by environmental factors during

Table 7.3 The dimensions of the optimized cyclone separators

Component	Barrel diameter	Barrel length	Cone length	Vortex diameter	Vortex length	Inlet diameter	Inlet height	Inlet width	Inlet position from the top	Outlet tip
Geometry	D_b	L_b	L_c	D_v	L_v	D_i	H	W	Z	D_o
Cyclone I, mm	320	420	750	120	250	100	125	95	70	100
Cyclone II, mm	320	420	750	120	250	100	125	95	0	100
Cyclone III, mm	320	420	750	120	250	100	195	95	0	100
Cyclone IV, mm	320	420	750	120	250	100	265	95	0	100
Cyclone V, mm	320	350	750	120	250	100	125	95	0	100

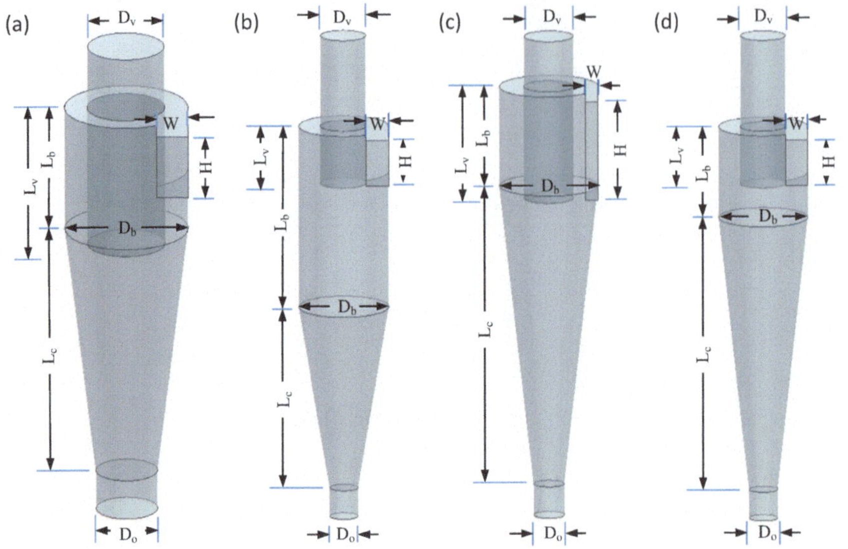

Fig. 7.3 Standard cyclones configuration; (**a**) 1D2D, (**b**) 2D2D, (**c**) 1D3D, (**d**) 1D3D w/2D2D inlet

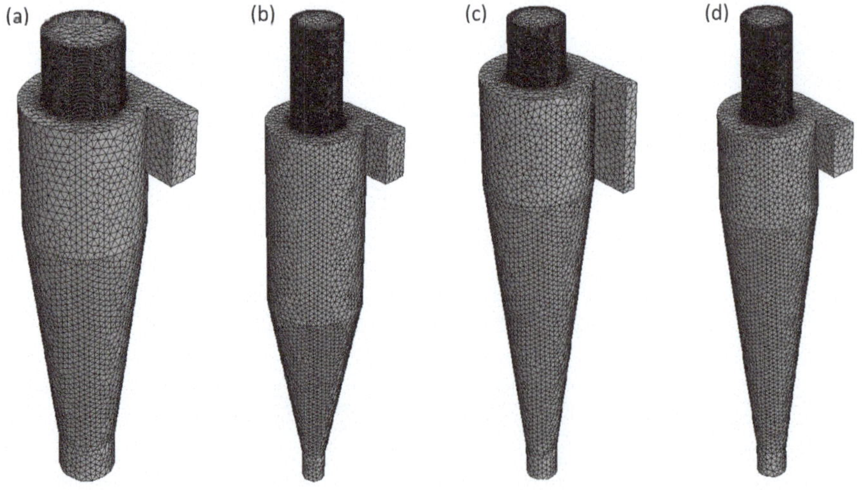

Fig. 7.4 Computational CFD grid mesh of the standard cyclones; (**a**) 1D2D, (**b**) 2D2D, (**c**) 1D3D, (**d**) 1D3D with 2D2D inlet

development, which can impact plant growth and yield [74]. On average, 1 kg of jojoba seeds contains approximately 800 to 1200 seeds, and their weight and shape classification ranges from small to medium to large. The annual seed production per bush is about 2–3 kg, which translates to 2–3 metric tons per hectare, depending on the male-to-female ratio in the existing plantations. The diverse physical

Table 7.4 Non-dimensionalized dimensions components of the standard cyclones

Component	Barrel diameter, D_b	Barrel length, L_b	Cone length, L_c	Vortex diameter, D_v	Vortex length, L_v	Inlet height, H	Inlet width, W	Outlet tip, D_o
1D2D	D_b	D_b	$2 \times D_b$	$D_b/1.6$	$9D_b/8$	$D_b/2$	$D_b/4$	$D_b/2$
2D2D	D_b	$2 \times D_b$	$2 \times D_b$	$D_b/2$	$5D_b/8$	$D_b/2$	$D_b/4$	$D_b/4$
1D3D	D_b	D_b	$3 \times D_b$	$D_b/2$	$9D_b/8$	D_b	$D_b/8$	$D_b/3$
1D3D w/2D2D	D_b	D_b	$3 \times D_b$	$D_b/2$	$5D_b/8$	$D_b/2$	$D_b/4$	$D_b/3$

Table 7.5 The total grid cells number of the standard cyclone separators

Cyclone type	1D2D	2D2D	1D3D	1D3D w/2D2D
Number of grid cells	606,208	486,824	468,094	516,421

Fig. 7.5 Jojoba particles; (**a**) leaves, (**b**) seeds, (**c**) particle mixture

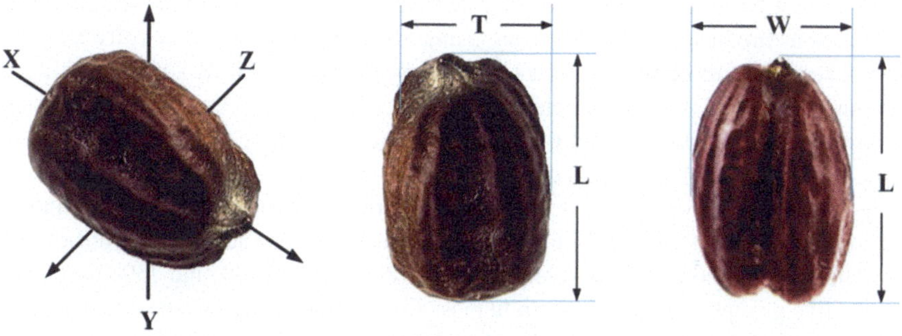

Fig. 7.6 Major dimensions of Jojoba seeds; X-axis is the length (L), Y-axis is the width (W), and the Z-axis is the thickness (T)

characteristics of the seeds make them a fascinating subject of study and play a crucial role in the cultivation and harvesting of jojoba plants. Jojoba seeds come in various sizes, ranging from 13.7 to 20 mm in length, 5.6 to 10.6 mm in width, and 5.2 to 13.2 mm in thickness, as illustrated in Fig. 7.6. Depending on the seed size, there can be anywhere from 1060 to 2300 seeds per kilogram [69]. These seeds have a density of 831 kg/m^3, with an equivalent diameter falling within the 9–12 mm range. To ensure optimal growth, it is recommended that the soil surface is suctioned at a velocity of 30 m/s, considering the seed size [59, 72].

When it comes to jojoba leaves, their unique characteristics cannot be ignored. Figure 7.7a displays the elliptical shape of these leaves, which measure between 20–40-mm long, 10–20-mm wide, and 0.05-mm thick and feature sharp apexes. These leaves are known for their thickness, waxy texture, and glaucous gray-green color. With an average density value of 500 kg/m^3, a terminal velocity of 11.5 m/s and an average projected area of 886 mm^2 [59, 75]; jojoba leaves are a marvel of nature. In the food processing and oilseed trading industry, separating and purifying cereal grains and seeds from impurities is crucial [76]. This process requires advanced technological procedures that consider each product's physio-mechanical characteristics. Impurities are eliminated based on size (width, length, and thickness) and drag coefficient [77]. Many equipment and installations are used to achieve this, including aerodynamic separators, screening devices, magnetic separators, vibrating separators, and more [78]. Herein, the current work uses aerodynamic systems, namely, aerocyclone separators, to separate cereal grains and seeds from impurities. These innovative systems have proven highly effective in the separation process, making them a popular choice in the industry.

In numerical simulations, agricultural grains and seeds are often assumed to have a symmetrical geometry, even though many of these biogenic particulates are not spherical. A more realistic approach in this simulation process is utilized to address this issue. As shown in Fig. 7.7b, we directly modeled the seed particulate matter using the DEM code, which allowed us to accurately represent their true geometry and dimensions as a sphere-polyhedron shape. The vertical and horizontal aspect ratios were 1.6 and 0.75, respectively. Similarly, we also modeled the leaves using their actual dimensions and geometry using UG graphics Nx-CAD software. We then saved these models as .stl files and exported them to DEM software using a custom polyhedron shape command. By factoring in the interaction of the particles during the simulation and setting their velocities to match the cyclone inlet air velocity, we were able to gain a deeper understanding of actual particle behavior.

In the first case, 500 Jojoba seed particles were modeled and successfully introduced into the cyclone inlet. While five sets of leaves were successfully introduced with the seeds into the cyclone inlet with 500, 1000, 1500, 2000, and 2500 particles. So, the five groups of biogenic particle mass loading introduced to the cyclone inlet contain the number of particles of 1000, 1500, 2000, 2500, and 3000 particles. At this level, the seeds:leaves ratios are considered to be approximately 0.1, 0.2, 0.3, 0.4, and 0.5 by weight depending on the density and physical properties of the

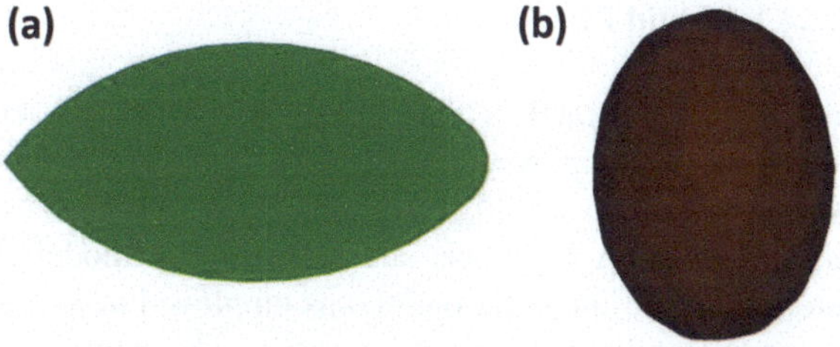

Fig. 7.7 The particle geometry modeling used in the simulations: (**a**) leaf, (**b**) seed

particles, respectively. In the second and third cases, the maximum particle mass loading is considered to optimize cyclone performance. At this level, 3000 particles are modeled and introduced into the cyclone's inlet as 500 particles of Jojoba seeds and 2500 particles of Jojoba leaves, and the seeds:leaves ratio is considered to be 0.5. During the simulation, the feed solids concentration is approximately 1.02 $kg_{air}/kg_{particle}$ according to the particles' physical properties, where the number of particles introduced to the cyclone inlet is 1500 per second. The seeds are assumed to be trapped by the bottom of the cyclone (product). However, the leaves are predicted to escape through the vortex finder tube (reject).

7.2.4 Granular Flow Modulation

In this numerical application, a CFD–DEM one-way coupling approach was utilized to simulate the granular flow in centrifugal reverse-flow cyclone separators accurately. This simulation involved air flowing through the cyclones, with the flow modeled to reflect the true shape of biogenic particles such as Jojoba seeds and leaves. While the simulations were not conducted under identical conditions as the experiment, the data analysis is still valuable, particularly qualitatively. The drag correlation model played a vital role in this coupling simulation, mainly when dealing with the interaction between the surrounding fluid and spherical and non-spherical particles and particle concentration. It is noteworthy that the one-way coupling approach was chosen for several reasons. First, the velocity of the particles at the inlet of the domain was the same as the air velocity [79–81]. Second, the effect of the particles on the fluid was considered negligible, enabling the steady-state simulation to be performed [82]. Third, the particles' relatively large size and density (10–30 mm) necessitated this approach [83]. Fourth, the Ganser drag model was used to identify different particle shapes, and this model is only available in one-way coupling [81]. Fifth, due to the large geometry size and true particle modeling, one-way coupling was the most time-efficient approach, taking significantly less time than other coupling methods, which could take weeks per simulation run.

7.2.4.1 Fluid Flow

In order to accurately simulate airflow in cyclone separators, cutting-edge CFD software was employed. Specifically, the finite volume CFD-based code using ANSYS Fluent® was utilized, with a double-precision solver for enhanced calculation accuracy. A Reynolds stress turbulence model (RSM) with 5% turbulence intensity at the inlet and outlet was employed to account for the airflow's chaotic and turbulent nature within the cyclones. This approach has been widely recognized as the most effective way to describe fluid flow behavior under such conditions, as confirmed by previous studies [84–88]. Throughout the simulation, the air properties were assumed to be turbulent, compressible, and isothermal, with a temperature of 25 °C, a density of 1.225 kg/m^3, and a kinematic viscosity of 1.7894×10^{-6} kg/m.s. A uniform air velocity distribution at 30 m/s was introduced at the inlet boundary condition, as recommended by several investigations [59, 72], for effectively capturing the targeted particles from soil surfaces. At the outlet, atmospheric pressure boundary conditions were utilized, while the no-slip boundary condition was assumed at the wall. To discretize the partial differential equation for modeling pressure-velocity coupling, the finite volume algorithm method was implemented using the SIMPLE (semi-implicit method for pressure-linked equations-consistent) scheme, with second-order upwind discretization for momentum, turbulent kinetic energy, and turbulent dissipation rate. The convergence criteria were set at 10^{-5} for continuity and 10^{-3} for other equations, with the resulting computational domains of the cyclones used to conceptualize velocity and pressure profiles. Following the CFD Fluent® simulations, the results were saved with the (.cas) extension, with the velocity and pressure results exported to DEM software as a (.f2r) formatted file. A comprehensive list of CFD numerical parameters used in this simulation is provided in Table 7.6. This cutting-edge methodology could accurately simulate and analyze airflow in cyclone separators, with significant implications for industries that rely on such technology.

Table 7.6 CFD numerical parameters used in simulating fluid flow

Parameter	Unit	Value
Inlet air velocity	m/s	30
Airflow rate	m^3/s	0.235
Air density	Kg/m^3	1.225
Air viscosity	Kg/m.s	1.7894×10^{-6}
Convergence criteria	–	10^{-3} to 10^{-5}
Turbulence model	RSM	–
Outlet	Pressure-out gauge (pa)	0
Wall	Standard non-slip	–
Pressure-velocity linking	SIMPLE	–
Turbulence intensity	–	5%

Table 7.7 Baseline of DEM particle parameter values

Parameter	Unit	Seed	Leaf
Particle shape	–	Sphero-polyhedron	Custom polyhedron
Dimensions	mm	11.64	30 × 15 × 0.05
Density	kg/m^3	831	500
Velocity	m/s	30	30
Number of particles	–	500	500–2500
Static and dynamic friction	–	0.3/0.3	0.3/0.3
Coefficient of restitution	–	0.9	0.8
Rolling resistance	–	0.3	0.3
Poisson's ratio	–	0.3	0.3
Young's modulus	GPa	100	10
		750 (boundary—particles contact)	
Drag model	–	Ganser drag model	

7.2.4.2 Particle Flow

Granular flow simulations require accurate modeling of normal and tangential contact forces to predict real-world results effectively. One promising software for this purpose is DEM Rocky®, a cutting-edge package developed by Granular Dynamics International, LLC, in collaboration with Engineering Simulation and Scientific Software Company (ESSS). Its shared-memory parallel architecture enables the rapid solution of granular mechanics problems, making it a potent tool in the commercial DEM world. One of DEM Rocky®'s unique capabilities is its true modeling of non-round particle shapes, allowing for more accurate simulation of real-world scenarios. Furthermore, the software can simulate particle breakage without loss of mass or volume and even account for shape changes on boundary surfaces due to wear. In the simulations conducted for this study, the latest version of DEM Rocky® was utilized, using two main biogenic particles, namely, Jojoba seeds and Jojoba leaves, to represent the solid phase. Unlike other DEM software that relies on clusters of spheres to model non-rounded particles, DEM Rocky® uses the particles' actual dimensions, geometries, and aspect ratios, providing more reliable results. Both normal and tangential interactions between particles were considered during the simulation, with a fixed velocity of 30 m/s set for the inlet air. Table 7.7 summarizes the specific properties of the particles used in the simulations. With its advanced capabilities and accuracy, DEM is poised to make significant contributions to the field of granular flow simulations.

The investigation of cyclones requires a detailed simulation of particle behavior. To achieve this, a (.cas) formatted file was exported to DEM, ensuring the geometric models for both DEM and ANSYS Fluent® packages were identical. The fluid flow results, including velocity and pressure profiles, were exported as (.f2r) extension files and integrated with DEM using one-way coupling processing. Given the particles' varied concentration, alignment, and shape, such as sphero-polyhedron and custom polyhedron, the Ganser correlation model [89] was expertly applied. To replicate real-world scenarios, particles were continuously fed into the cyclone inlets for the first 2 s of the 6-s simulation period, while airflow was continuously supplied throughout. The one-way coupling simulation of the DEM and ANSYS Fluent® computations was highly complex and time-consuming, requiring between 48 and 130 h per run depending on the particle mass loading and cyclone geometry. Ultimately, 13 DEM and ANSYS Fluent® simulation coupling runs were performed, with each model run having its unique parameters. To accurately replicate the behavior of particles within the cyclones, it was assumed that the seeds were trapped at the bottom of the cyclone (product), while the leaves escaped through the vortex finder tube located at the top center portion of the cyclone (reject). With these highly sophisticated simulations, the researchers were able to gain insights into the complex behavior of cyclones, enabling them to optimize their design and performance in a highly efficient and effective manner.

Friction, a term used to describe the resistance between two surfaces in contact, is an essential component of understanding the physical properties of materials. In simple terms, when two surfaces come into contact, friction is generated due to their relative motion. If there is no slipping, the friction is known as static, while if slipping is present, it is called dynamic friction. Making certain assumptions to achieve accurate results while maintaining fast processing is crucial in simulations. For instance, the surface of the particles is assumed to be smooth and spherical, and their friction value and rolling resistance must be no lower than 0.3, as recommended [90]. To simplify the analysis process and reduce the number of variables, the static and dynamic friction values for both particle–particle and particle–wall interactions can be assumed equal [83, 91–95]. Additionally, according to DEM technical manual notifications [90], Young's modulus of materials can be reduced to 1×10^6 to 1×10^7 Pa. without significantly affecting the simulation results. While a higher value of Young's modulus would lead to more accurate results, it would also increase processing time. Similarly, the Poisson's ratio value is set to the program default [91, 92]. The coefficient of restitution, which measures the energy dissipation during a collision, is set to a high value in this simulation due to the hardness of the particles [96–99]. This value ensures that the collision between particles is as elastic as possible, reducing the amount of energy lost during the interaction. In conclusion, understanding the role of friction and making certain assumptions can significantly improve the accuracy and speed of simulations, leading to a more effective and efficient analysis of material properties.

7.2.5 Experimental Measurements

The economic value of jojoba seeds is undeniable due to their precious oil content. Therefore, preserving and harvesting them with the utmost quality is crucial. A cyclone separator is a reliable tool for achieving this goal. The previous study [59] provides detailed information on harvesting jojoba seeds. Cyclone effectiveness is a crucial indicator of the quality of this technique's harvesting process. This investigation aims to optimize the cyclone geometry to predict high performance in separating and cleaning efficiencies of jojoba seeds. To achieve optimal cyclone performance, the separation and cleaning efficiencies of the cyclone must be maximized. Separation efficiency is defined as the number of seeds collected from the granular flow, while cleaning efficiency refers to the quality of collecting these seeds without any undesired materials [100–102]. By introducing a known number of seeds and leaves to the cyclone inlet, simulations can determine the amount of each particle type that emerges from the outlet. Using this information, the separation efficiency, Eq. (7.1), can be calculated by comparing the number of seeds collected in the outlet with the number of seeds introduced to the cyclone inlet. The seed loss percentage, represented by the number of seeds that are not collected in the product portion, can be predicted using Eq. (7.2). Similarly, the cleaning efficiency, Eq. (7.3), can be calculated by analyzing the ratio of seeds to the total amount of particles in the outlet. Effectiveness is a crucial parameter that provides insight into the efficiency of the cyclone process. It tracks improvements in the separation process over time, focusing on two key parameters: cyclone performance (separation efficiency) and product quality (number of rejected leaves). This can be formulated as an Eq. (7.4) [103, 104]. During the simulation, particle mass loading ratios were converted to particle numbers based on their density and weight parameters. Optimizing the cyclone geometry can lead to high separation and cleaning efficiencies of jojoba seeds. This, in turn, maximizes the economic value of jojoba seeds by preserving their precious oil content. Using effectiveness as a parameter, it can track improvements in the separation process and ensure that the cyclone process runs effectively over time.

$$\eta_s = \frac{\text{Seeds number in product}}{\text{Seeds number in feed}} \times 100 \tag{7.1}$$

$$S_l = 100 - \eta_s \tag{7.2}$$

$$\eta_c = \frac{\text{Seeds number in product}}{\text{Seeds number in product} + \text{Leaves number in the domain}} \times 100 \tag{7.3}$$

$$\eta_{\text{eff}} = \frac{\eta_c \times \text{Leaves separated ratio}(\text{reject})}{100} \tag{7.4}$$

where η_s, S_l, η_{eff}, and η_c are the separation efficiency, seed loss, cyclone effectiveness, and cleaning efficiencies, %, respectively.

7.3 Results and Discussion

In this section, the results of the simulations are compared and validated with the experimental data and other investigations. The comparison is only qualitative since the present simulations were not performed under the same conditions as the experiment environment.

7.3.1 *Model Validation*

Efficient particle separation is crucial when using gas cyclones, as it ensures the removal of undesired materials. To attain accurate simulation results, it is imperative to validate them by comparing them with experimental data. This validation process involves a step-by-step investigation, comparing measurements from previous empirical studies with CFD–DEM one-way coupling simulations of a specially designed cyclone [59]. To achieve this, we conducted tests on a jojoba seed harvesting machine using a cyclone with a main body diameter of 320 mm, as depicted in Fig. 7.8, at the Agricultural Engineering Department Laboratory in Alexandria, Egypt. For more comprehensive details on the jojoba harvesting methodology, please refer to the relevant literature [59, 72, 105, 106].

The designed cyclone geometry allowed for air velocity to be generated from a top-mounted blower, as it was designed. However, the fan was removed during simulations; only the flow's impact was factored in. To simplify the evaluation of the cyclone's performance during simulation, the particle mass loading ratios from the experiment were converted to particle numbers based on their density and weight

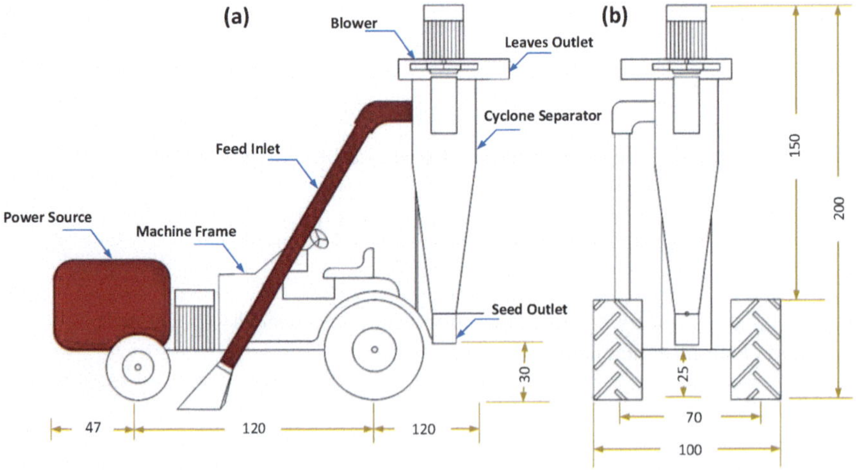

Fig. 7.8 Schematic line drawing of the jojoba seed harvesting machine: (**a**) side view, (**b**) back view; all dimensions in cm [59]

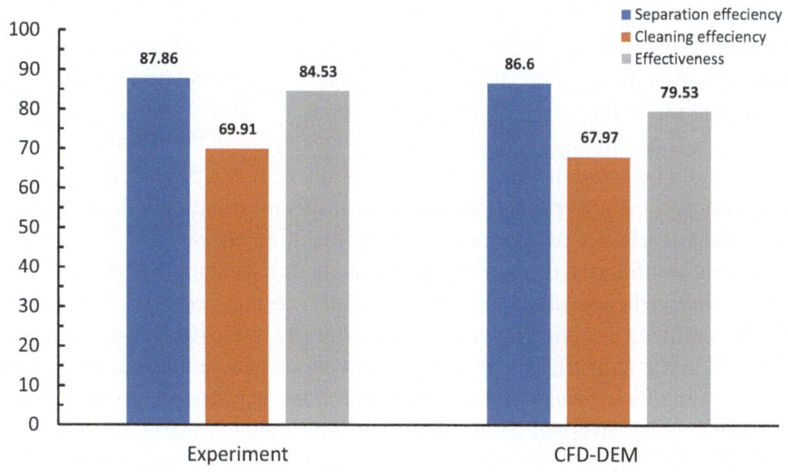

Fig. 7.9 Comparison between the experimental and the simulation result data of the designed cyclone at a ratio of 0.5

parameters. Five weight ratios of seeds to leaves were employed in the experiment (0.1, 0.2, 0.3, 0.4, and 0.5) to assess the cyclone's efficiency, chosen randomly from the farm's harvest season samples to provide a representative range of ratios. The first simulation case utilized the separation efficiency parameter to validate the simulation results for different loading ratios. As shown in Fig. 7.9, the results demonstrated excellent agreement between the simulation and experimental data, with a minimal average divergence of 0.23% for the first three ratios and 1.54% for the last two ratios, indicating the model's capability to predict cyclone performance. In the second and third simulation cases, with a loading ratio of 0.5, the goodness of fit between the designed cyclone's experimental data and numerical results was deemed highly acceptable, as depicted in Fig. 7.9. However, no experimental data were available to validate the optimized cyclones' results, including II, III, IV, and V, and the standard cyclones 1D2D, 2D2D, 1D3D, and 1D3Dw/2D2D inlet. Nonetheless, simulations of the optimized cyclones, specifically (III) and 1D2D, showed better performance than the numerical outcome of the designed cyclone. Thus, the present simulation methodology is progressing in the right direction toward predicting the cyclone's performance, utilizing distinct particle geometries.

Fortunately, the designed cyclone has shown promising results in separating efficiency after thorough simulations. The deviation value of a mere 1.26% from experimental data is a testament to the accuracy of the simulation. The cleaning efficiency calculation also showed a small discrepancy of only 1.94% less than the experimental results, indicating the cyclone's effectiveness. However, it is worth noting that the cyclone's effectiveness decreased by 6.3% due to its reliance on separation and cleaning efficiencies. Upon closer examination of the simulation results, it was discovered that most leaves were successfully separated but remained trapped within the top section of the cylindrical area. This accumulation of debris can negatively impact the cyclone's performance during continuous operation. The issue was traced back to the inlet portion's design, which was constructed to be 70 mm below the top surface of the cyclone. To improve the cyclone's overall performance, attention has been focused on modifying the inlet geometry. The objective is to optimize the cyclone's functionality, as discussed in greater detail in the following sections. While it should be noted that the simulation results were not produced under identical conditions to the experiment, the analysis of this information will be beneficial, at least qualitatively, for future design and optimization endeavors.

To ensure confidence in the simulation methodology and validate the contact model, mainly when dealing with particles of varying geometries and shapes, researchers utilized a mixing tee and wind shifter [92, 107, 108], two devices commonly used in the industrial waste process to separate heavier particles from lighter ones. Following their results and workflow for a one-way coupling simulation, the team conducted a converged steady-state fluid flow analysis using ANSYS Fluent® software. Then it exported the result to DEM for further analysis. To simulate particle movement, the Ganser correlation drag law was deemed the most suitable option, as it considers particle geometry, which significantly impacts drag force and can lead to more accurate results when separating particles. The phenomenon of particle trajectory within cyclone separators was then investigated. It confirmed that the seeds could reach a macroscopic steady-state at the bottom of the conic part within 2 s of simulation time and could be accurately captured by their numerical model [61, 109]. Ultimately, it was determined that this simulation methodology could be confidently adopted for other cyclone simulations, leading to a better understanding of the most effective design for gas cyclone separators when separating and cleaning jojoba seeds. The results of this study have significant implications for the industrial waste process and offer exciting opportunities for future research in this field.

7.3.2 Computation Case One (Designed Cyclone)

7.3.2.1 Fluid Profile

As a valuable tool in fluid dynamics, CFD numerical simulations provide critical insights into the behavior of cyclone separators, especially when understanding the velocity and pressure distribution effects of fluid flow, which are notoriously difficult to measure experimentally. CFD simulations can predict and analyze the airflow behavior inside a designed cyclone separator based on the velocity and pressure distribution fields, as depicted in Fig. 7.10. By examining the pressure fluctuation levels across the cyclone domain (as illustrated in Fig. 7.10b), it becomes evident that pressure is highest near the cyclone wall, gradually decreasing in the radial direction and ultimately becoming negative at the center of the cyclone. Notably, the maximum negative pressure zone is located at the beginning of the vortex finder in

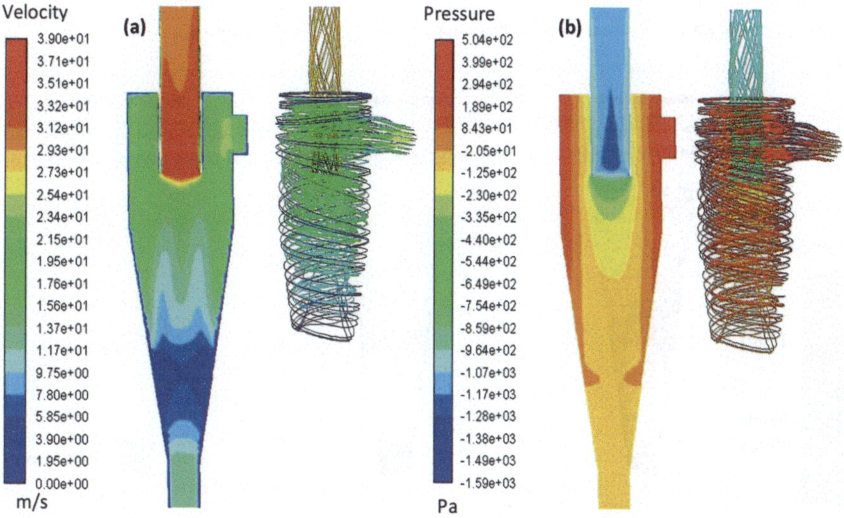

Fig. 7.10 Airflow profile with streamlines inside the designed cyclone separator at an inlet velocity of 30 m/s: (**a**) velocity, (**b**) pressure

the top center of the cyclone. Additionally, as the length of the cyclone increases, the length of the negative pressure zone decreases, suggesting that airflow in this region will have a more significant impact on particles than in other areas. Then, it becomes positive in the middle of the conical section and decreases gradually to be negative again at the end of the cyclone. This trend could indicate that the airflow will have a higher impact on the particles in this region than on others. At the end of the negative pressure zone, it is clear that the flow streamline reverses its direction toward the cyclone's center and escapes from the vortex finder. Notably, an accurate prediction of the velocity fluctuation levels is needed to obtain the correct result for the cyclone performance and show the fluid flow's swirling behavior. The velocity profile describes the characteristic flow structure known as the Rankine vortex type [110–112]. This vortex structure is famous for flow inside the reverse flow cyclone separators. It consists of an outer free vortex and a forced inner vortex at the center. The flow near the cyclone walls is directed to push the heavy particles toward the bottom outlet tip of the cyclone. At the conic section, the fluid flow direction (velocity streamline) is reversed upwards in the central region via the vortex finder toward the cyclone outflow, as shown in Figure 7.10a. In the lower region of the conic part (the seed outlet portion), the air velocity is lower than the terminal velocity of the seeds, which will yield a good indicator that the seeds will be halted and come out from the bottom of the cyclone.

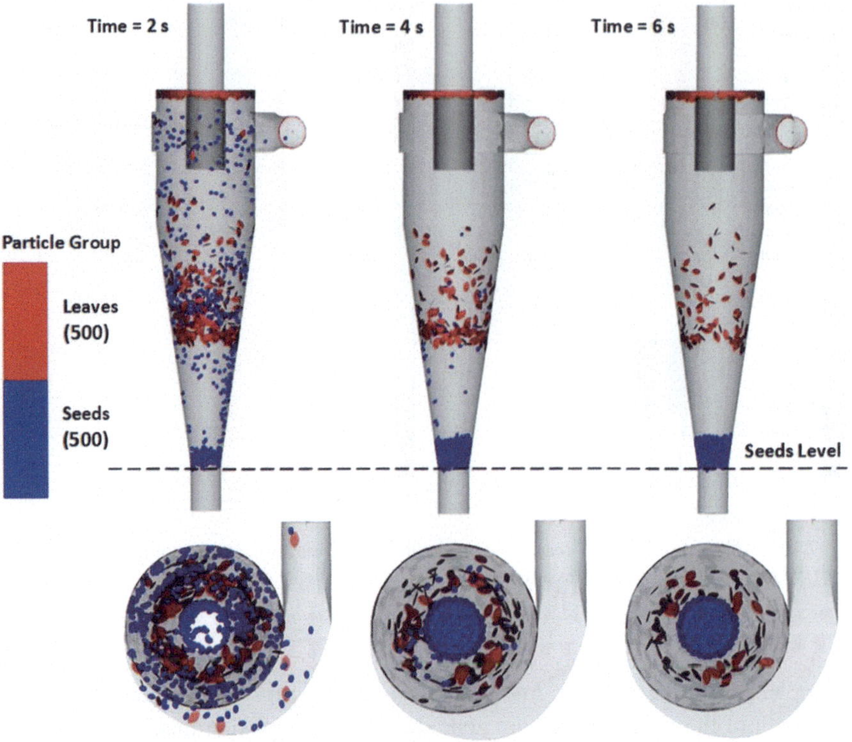

Fig. 7.11 Particle trajectories during the simulation at different simulation times (ratio = 0.1)

7.3.2.2 Particle Trajectory

In cyclone separators, the behavior of macroscopic flow particles is of utmost importance. These particles tend to amass at the wall immediately after entering the cyclone inlet, eventually descending in strands or bands [113, 114]. This pattern has proven difficult to predict accurately in past numerical models. However, recent CFD–DEM modeling developments have successfully captured this phenomenon [115, 116]. As demonstrated in Figs. 7.11 and 7.12, the CFD–DEM model can depict the macroscopically steady-state flow achieved after approximately 2 s, in agreement with previous research [61]. Strands or bands' flow pattern indicates the cyclone's separation efficiency. During simulation, supporting forces act on the particles adjacent to the conic wall of the cyclone, leading to a higher descending velocity in the cylindrical portion than the conical portion [114]. At a particle ratio of 0.5, the simulation results shown in Fig. 7.11 indicate that the particle flow enters the cyclone inlet and spirals around the barrel portion near the top before descending tangentially into the conic portion, creating an outer vortex. The air velocity within this vortex increases, leading to a greater centrifugal force that separates and congregates the particles near the wall. As the flow reaches the conic portion, the air

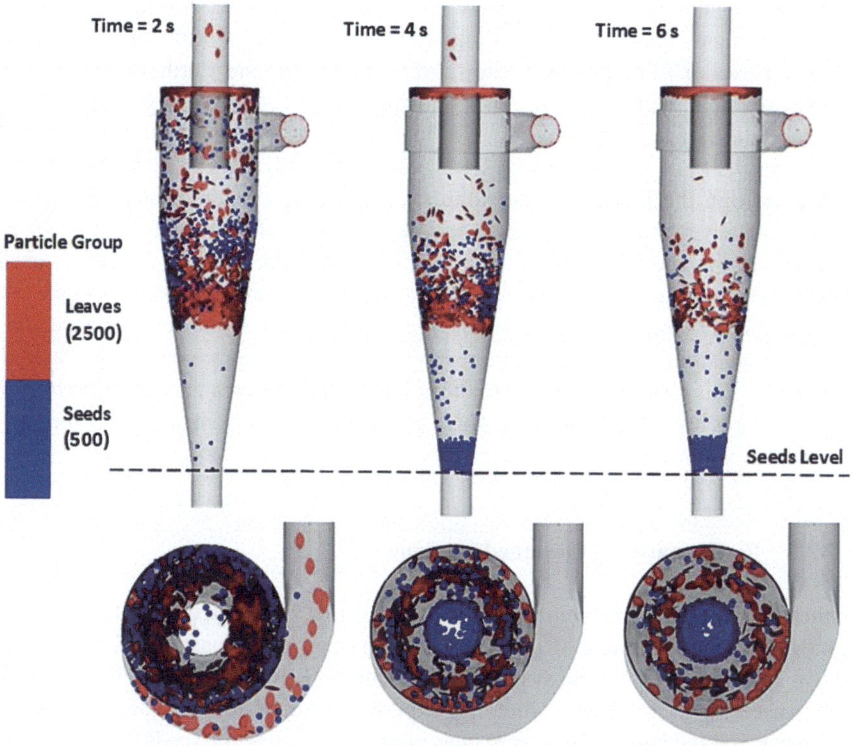

Fig. 7.12 Particle trajectories during the simulation at different simulation times (ratio = 0.5)

velocity drops precipitously, resulting in an inner vortex and a reversal of the air stream toward the top center of the cyclone. The leaves are dispersed throughout the cyclone and are likely to be caught in the flow through the vortex finder, while the seeds accumulate in the wall region and gradually move toward the seed outlet.

Considering the complexity of the turbulent swirling flow pattern of macroscopic particles inside the investigated cyclone separator, the snapshots in Figs. 7.11 and 7.12 at seed:leaf ratios of 0.1 and 0.5, respectively, made the particles separation process visual and clear for the one-way coupled simulation. Herein, the snapshots for these two ratios have been taken to show the clear difference during the separation process with a low and high concentration of particle loading. The leaves (light particles) are dispersed by turbulence throughout the cyclone and are likely to be caught in the flow through the vortex finder at the top center of the cyclone. The heavier the particles (seeds) are, the more they will accumulate in the wall region and gradually move (due to the combined action of the gas flow and gravity) to the seed outlet. These two phenomena are matched and captured by previous studies using a one-way and two-way coupling technique [117] and the Lagrange approach [85] using different particles with different diameters.

A significant amount of leaves were still inside the cyclone, despite being separated from the cyclone's body. This may be due to the design of the inlet, which is fixed at a level of 70 mm below the top surface of the cyclone. Further optimization of the cyclone's geometry is necessary to improve the separation process. When comparing the trajectories of particles at low and high loading ratios, it is apparent from Figs. 7.11 and 7.12 that many seeds reach the steady state at the bottom of the cyclone within the first 2 s of the simulation time. At a loading ratio of 0.1, the smallest number of remaining seeds not separated inside the cyclone domain is observed at the end of the simulation time. This finding indicates that most seed particles were separated from the leaves during the cyclonic process. However, as the leaf loading number increases, many seeds accumulate in the conic part, requiring increased simulation time and more intensive power usage to achieve the desired separation. This result is attributed to the heightened collision and interaction forces between particles in the conic part, which prevent the seeds from settling at the bottom of the cyclone.

7.3.2.3 Performance of Designed Cyclone

In the simulation, five different particle loading ratios were tested (0.1, 0.2, 0.3, 0.4, and 0.5), while the number of seeds was kept constant at 500 particles. The number of leaves introduced into the system varied, with 500, 1000, 1500, 2000, and 2500 leaves used for each loading ratio. The experiment aimed to assess the separation efficiency and seed loss, which were measured by collecting the seeds that fell to the bottom of the conical section of the cyclone. The results in Fig. 7.13 show that the number of collected seeds varied significantly depending on the leaves injection ratio. Specifically, as the loading ratio increased from 0.1 to 0.5, the collected seeds decreased from 492 to 433. These findings suggest that increasing the leaves loading ratio led to a decrease in the separation efficiency and an increase in seed loss. Notably, at the end of the simulation time step, the remaining seeds inside the

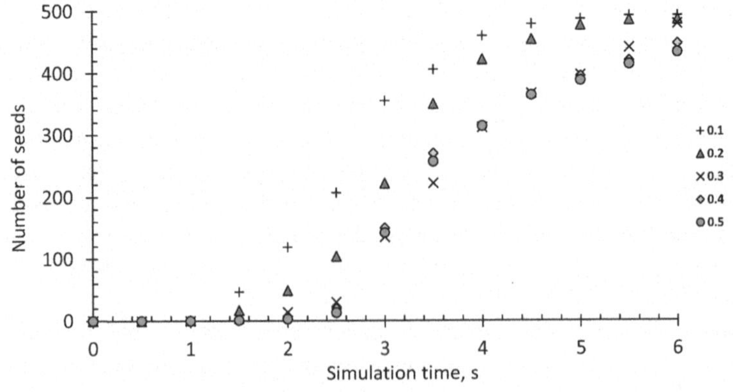

Fig. 7.13 The number of seeds collected during the simulation process

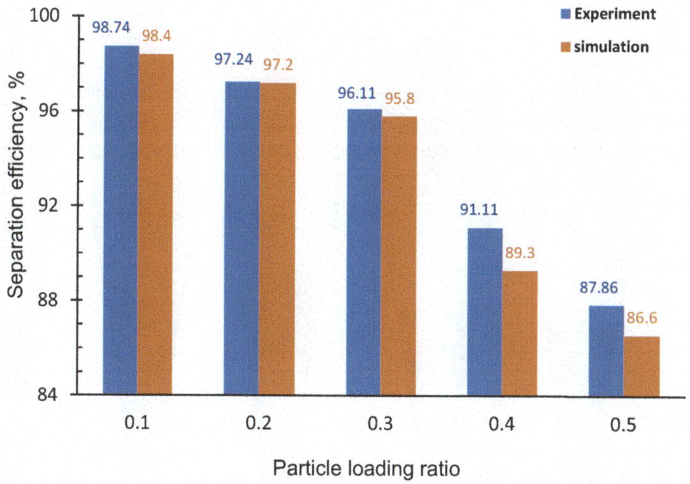

Fig. 7.14 Separation efficiency comparison between the simulation and the experimental results

cyclone domain that were not collected at the bottom were neglected from the calculations. The seed collection trends during the simulations have the same behavior; however, increasing the leaves loading ratio causes the number of collected seeds to decrease gradually. The reason why the number of collected seeds at the bottom of the cyclone is lower at higher leaves loading is that the particles tend to keep rotating in the conic section when the flow is dense, and the particle-particle interaction force is large enough to prevent the particle from moving downward quickly.

The cyclone separation efficiency and seed losses can be easily calculated by considering the number of separated seeds accumulated at the bottom of the cyclone using Eqs. (7.1) and (7.2), as explained in the previous section. Figure 7.14 was then generated using Eq. (7.1), which displays the total separation efficiencies at different particle loading ratios. Both simulation and experimental data were presented in the figure, and they show good agreement. However, there is a slight average deviation of 0.23% for the first three ratios and 1.54% for the last two ratios between the simulation and experimental results. This indicates that the proposed model can predict cyclone performance outcomes accurately. It is possible that the assumptions made during the construction and simulation of the model contributed to this slight deviation in the simulation results.

As the particle loading ratio increased in the simulation, the separation efficiency decreased and reached a minimum value of 86.6%. This was accompanied by an increase in seed losses, which were recorded at 1.6%, 2.8%, 4.2%, 10.7%, and 13.4% as the particle loading ratio gradually increased. The reason behind this trend was the movement of particles in the domain, where interactions between particles and with the walls caused the seeds to escape in the same direction as the leaves and thus exit through the vortex finder.

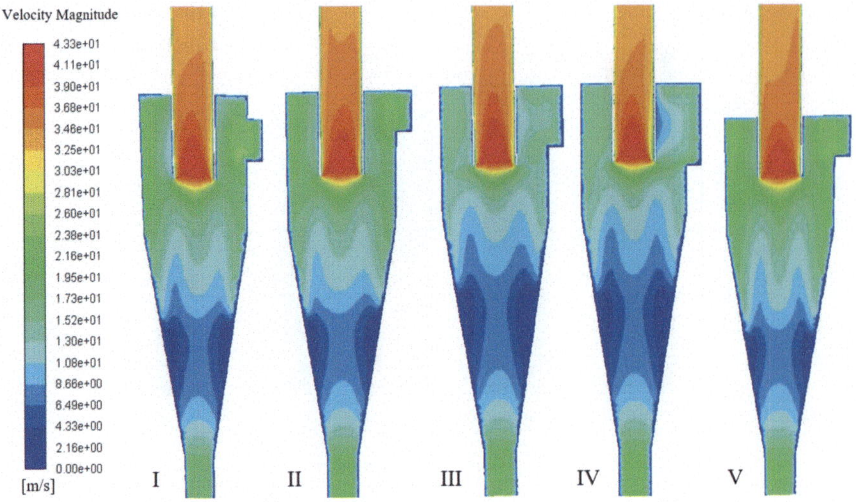

Fig. 7.15 Velocity profile distribution for the four optimized cyclones with the designed one

7.3.3 Computation Case Two (Optimized Cyclones)

7.3.3.1 Fluid Profile

In the realm of particle separation, the chosen inlet air velocity of approximately 30 m/s is an essential factor in achieving optimal results. Previous literature recommendations have highlighted this fact [72], and further research has proven it vital. By carefully examining Fig. 7.15, it can be observed that the air average velocity profile as it passes through the optimized cyclones. Intriguingly, a clear trend emerges during the simulations where the high-velocity value of the inner flow structures is recorded at the vortex finder's beginning. Meanwhile, at the bottom of the cyclone, the air velocity was recorded as less than the terminal velocity of the seeds. The conical section, however, represents the lowest point for the air velocity. These unique phenomena separate light particles, such as leaves, from heavier particles, like seeds. Utilizing the vortex finder, leaves can pass through unobstructed, while the seeds are led toward the bottom of the cyclone, resulting in highly efficient separation.

7.3.3.2 Particle Trajectory

Based on previous simulations and research data [113], a cyclone separator's design significantly impacts particles' movement and pathlines within its domain. The findings, depicted in Fig. 7.16, reveal that the trajectories of leaves in cyclones I and IV are concentrated at the top of the barrel section. In contrast, cyclones II and V exhibit a concentration of leaves in the conical section with the lowest air velocity

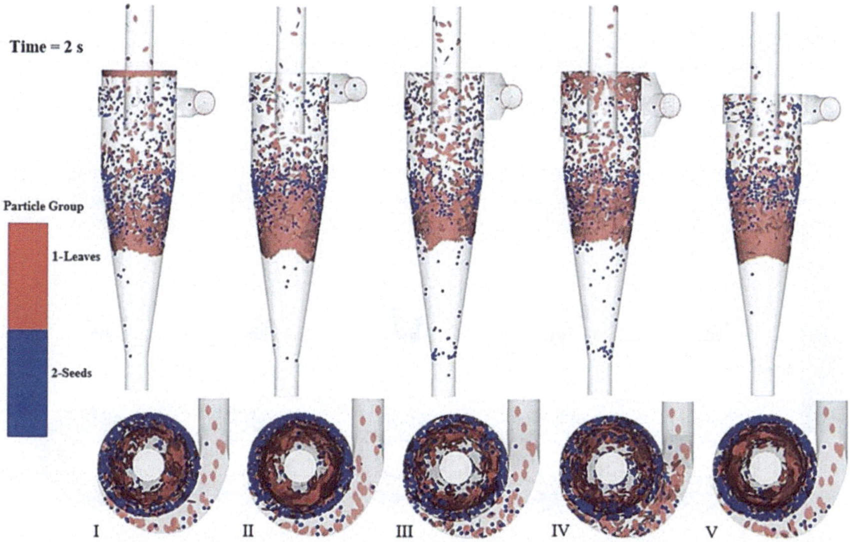

Fig. 7.16 Realizations of the particles' trajectories in the elevation and plane cross-section during DEM simulation at 2 s

profile. Cyclone III, with its regular distribution of leaves, facilitates the swift separation of swirls, allowing for the efficient separation of seeds. This is in contrast to the other studied cyclones, where a large number of leaves remain with the seeds for a more extended period, affecting cyclone performance. Regarding seed trajectories, cyclone III presents the highest number of collected seeds (product) and achieves a macroscopically steady-state flow at the bottom of the cyclone after 2 s of simulation time. These DEM simulations validate previous findings [61] and demonstrate that the concentration of seeds increases near the wall, forming an outer vortex, while the leaves are separated from the seeds, forming an inner vortex near the center of the cyclone. When introducing particles to the inlet, all cyclones show similar tendencies, except for cyclone IV, where a large number of particles remain in the volute inlet portion due to the inlet dimensions being relatively large. As a result, this consumes more simulation time, ultimately affecting cyclone performance. Our investigation into the theoretical operation conditions of the reverse cyclone, from the center to the wall boundary in the cross-section plane, further supports the importance of the cyclone's design. Differences in centrifugal forces on particles of varying densities lead to a concentration of seeds near the wall, forming an outer vortex, and the separation of leaves from seeds, forming an inner vortex near the center of the cyclone.

Simulation time is critical in determining the trajectory and concentration of leaves. As depicted in Fig. 7.17, at a simulation time of 4 s, the elevation cross-section highlights the leaves' trajectories within the cyclone designs. Notably, in cyclone (I), the leaves are predominantly concentrated in the top section, while

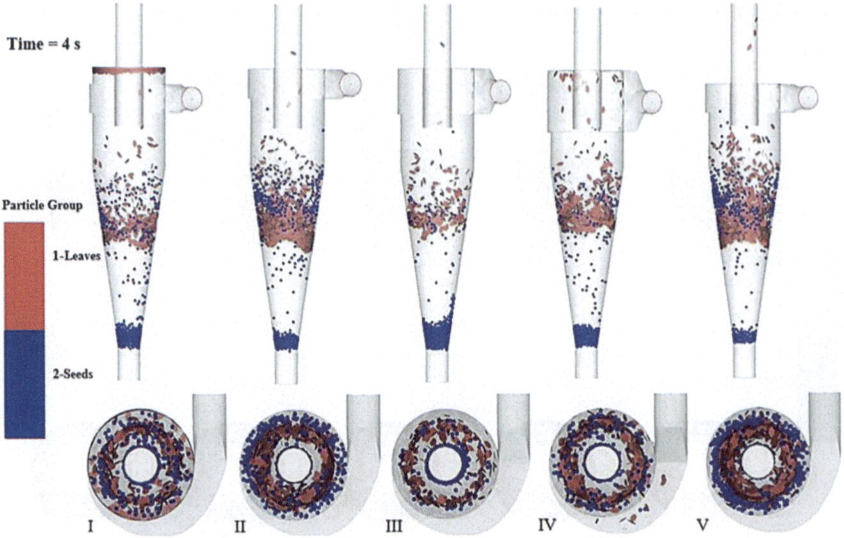

Fig. 7.17 Realizations of the particles' trajectories in the elevation and plane cross-section during DEM simulations at a 4 s

cyclones (II, IV, and V) exhibit intense concentration in the conical part. Comparatively, the number of leaves in cyclone (III) is significantly reduced due to their escape through the vortex finder, leading to improved cyclone performance. Moreover, cyclone (III) demonstrates a higher number of seeds reaching the steady-state flow in the product outlet than other designs, with cyclone (V) showing the lowest number. Examining the cross-section plane in Fig. 7.17, the particle vortexes observed in cyclones (II, IV, and V) remain heavy and follow the same path as the simulation time of 2 s. However, cyclone (III) presents a different scenario, with a considerable number of particles still in the volute inlet portion of the cyclone (IV). Further simulation time is required to move these particles into the barrel section, thereby enhancing the cyclone's effectiveness.

Finally, at the end of simulation time (6 s), it is crucial to consider the number of leaves remaining in the cyclone's domain and the number of seeds that have escaped from the product portion. These factors can significantly impact the separation and cleanliness efficiencies, which are critical in optimizing cyclone performance. Upon examining the elevation and plane sections in Fig. 7.18, it becomes evident that the number of leaves in the designed cyclone (I) is relatively high at the top portion. This could potentially result in decreased cyclone performance during continuous operation, making it essential to optimize the cyclone's design to achieve optimal results, which is precisely the aim of this study. Interestingly, the modified cyclones (II, V) demonstrate an unexpected outcome, with a large number of leaves still present in the conical part, where a heavy and robust vortex exists. This factor

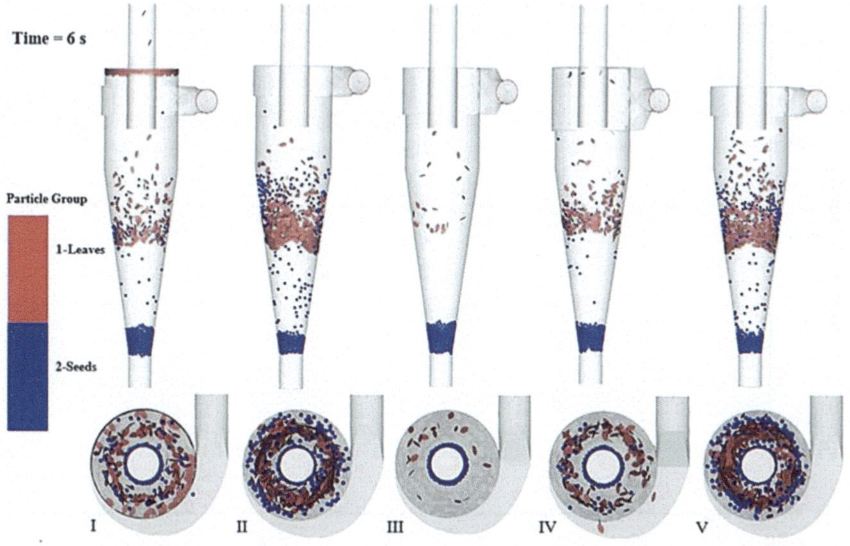

Fig. 7.18 Realizations of the particles' trajectories in the elevation and plane cross-section during DEM simulations at 6 s

can prolong the separation process, making it more time-consuming to dispose of the leaves from the reject portion of the cyclone, negatively affecting its effectiveness. A comparative observation reveals that the modified cyclone (III) performs exceptionally well, with all seeds reaching the cyclone's bottom in a steady-state condition. At the same time, only a negligible number of leaves remain in the domain, which has a minimal impact on the cyclone's performance. However, it is essential to note that some modified cyclones require a substantial simulation time to achieve this condition, reflecting the reality of high-power consumption and operation costs.

7.3.3.3 Performance of Optimized Cyclones

The effectiveness of modified cyclones can be evaluated through several crucial parameters, including separation efficiency, cleaning efficiency, and overall cyclone effectiveness. In order to determine cyclone performance for various simulation conditions and modifications to cyclone geometry, one can calculate the number of particles introduced to the inlet and the number of seeds and leaves that escape through the top and bottom outlets of the cyclone, respectively. Any seeds that remain within the cyclone at the end of the simulation can be neglected. With a simulation of 500 seeds introduced to the inlet, Fig. 7.19 illustrates the number of separated seeds collected in the cyclone's bottom portion. Remarkably, cyclones (III and IV) demonstrated a strong trend of surpassing the designed cyclone

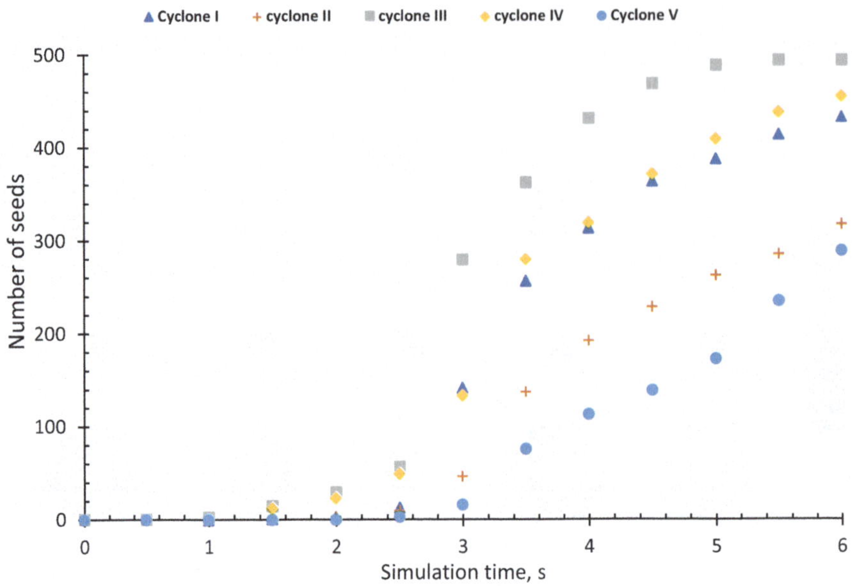

Fig. 7.19 Number of collected seeds in the product portion over simulation time

performance (I) by separating 493, 456, and 433 seeds, respectively. Meanwhile, cyclones (II and V) underperformed by separating only 318 and 289 seeds. Cyclones (III and IV) were the first to collect seed in the bottom portion, starting from 1 s, while the others began at 2 s. This highlights the critical impact of the cyclone inlet dimensions on performance, particularly in terms of separating seeds from the air-granular flow, more than modifications to the inlet portion and length of the cyclone barrel. These findings align with previous studies [60]. Notably, the curve tendency of the cyclone (III) reached stability at a simulation time of 5.2 s, while the relationships in the other cyclones still exhibited direct fit tendencies with simulation time, necessitating additional time to achieve a stable state. This could potentially affect cyclone effectiveness and increase power consumption during actual operation.

In the realm of seed collection, the purity of the harvest is of utmost importance. It is determined by the number of leaves that manage to elude the forces of the cyclone and escape through the vortex finder. This research delved into this phenomenon by simulating the separation of 2500 leaf particles within different cyclones, where the number of separated leaves during the simulation time step is illustrated in Fig. 7.20. Remarkably, the data yielded similar tendencies in cyclones (II) and V, where their performance in leaf separation was below average. In contrast, cyclones (I) and IV exhibited exceptional separation capabilities, separating and removing leaves early on. This trend was consistent across these two cyclones. Although cyclone (I) demonstrated good separation tendencies, it was discovered that a significant number of leaves remained trapped at the top of the barrel. This is

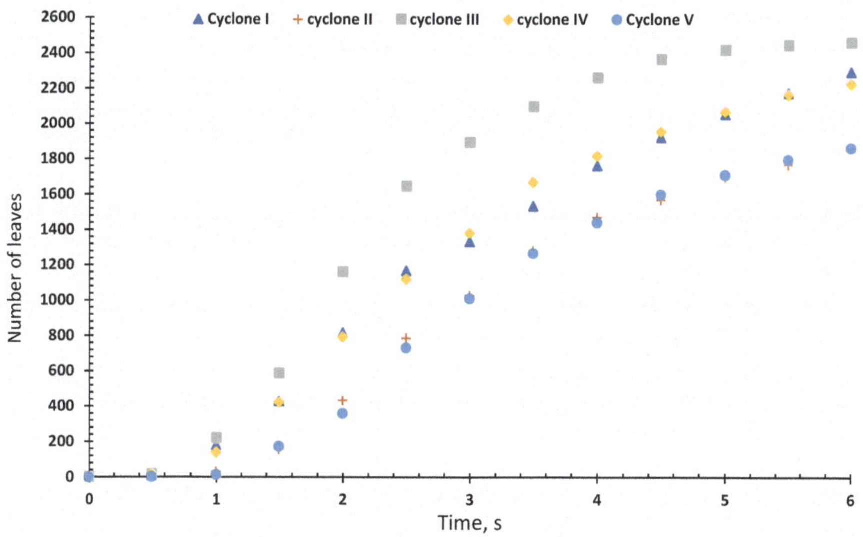

Fig. 7.20 Number of separated leaves during the simulation time

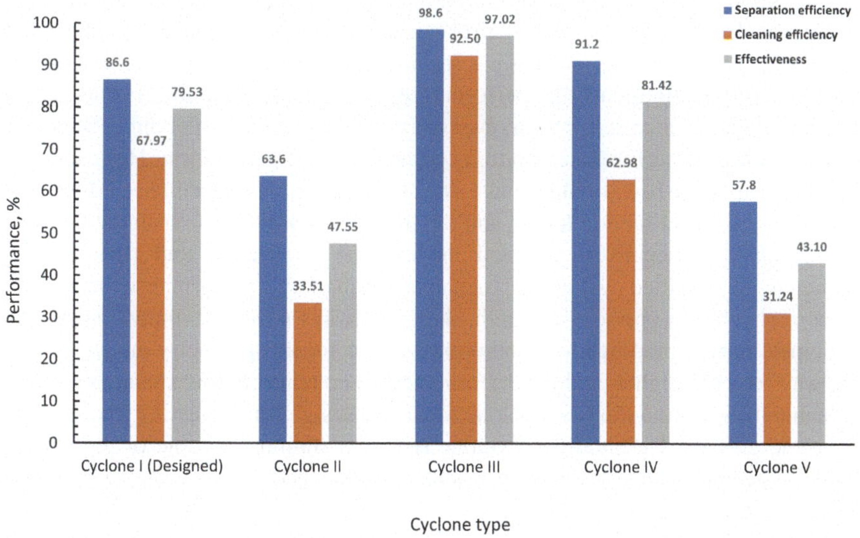

Fig. 7.21 Performance comparison of the modified and designed cyclones

attributed to the inlet portion's design, which is situated 70 mm below the top surface of the cyclone. As demonstrated in Fig. 7.20, cyclone III outperformed the rest in separating leaves, with a total of 2460, followed by cyclones I, IV, II, and V, separating 2296, 2232, 1869, and 1864 leaves, respectively.

The preceding sections demonstrated the simplicity of using Eqs. (7.1), (7.2), (7.3), and (7.4) to compute and visualize cyclone performance, as illustrated in Fig. 7.21. Based on empirical data, the separation efficiency, a crucial metric, achieved a commendable 87.86%, while the cleaning efficiency, which significantly impacts the efficacy of the machine, posted an impressive 69.91%, resulting in a machine effectiveness score of 84.53 [59]. Notably, during the modeling and simulation of the designed geometry cyclone (I), we observed that the cyclone performance matched the experimental data remarkably well, with deviations of only 1.26% and 1.94% for the separation and cleaning efficiencies, respectively. Consequently, the effectiveness of the cyclone, which depends on the separation and cleaning efficiencies, dipped by a marginal 6.3%.

In the quest for efficient and effective cyclone separators for harvesting jojoba seeds, the modified cyclone (III) reigns supreme. Outperforming its peers in geometric configuration and modification, it boasts a remarkable 98.6% separating efficiency, 92.5% cleaning efficiency, and 97.02% effectiveness. Its success is no coincidence, as it has been crafted to perfection, resulting in a higher performance that leaves other cyclones in the dust. While the modified cyclone (IV) has also proven to be a strong contender, achieving an impressive 91.2%, 62.98%, and 81.42% separating efficiency, cleaning efficiency, and effectiveness, respectively, it pales in comparison to its predecessor. On the other hand, the remaining cyclones (II, V) showed poor results, highlighting the negative impact that altering the inlet position and reducing the cyclone barrel can have on its performance. It is important to note that several factors, such as gas specifications, particle properties, and operational conditions, influence the performance of a cyclone separator. For large and true particulate matter, it has been observed that an increase in inlet dimensions can produce a higher performance up to a certain level of modification. However, further increases have been found to have a negative impact on performance. Moreover, changing the barrel section length and the inlet position without changing the inlet dimensions has negatively affected the cyclone's operation and performance. Although the modified cyclone (III) achieved a remarkable performance that has yet to be replicated in previous research, there is always room for improvement. We anticipate further modifications and optimizations for this geometry model, paving the way for even higher cyclone performance. In conclusion, this simulation offers valuable insights into the best cyclone separator design for harvesting jojoba seeds from the ground. The modified cyclone (III) is a shining example of the perfect blend of geometry and optimization, yielding unmatched performance and separating efficiency.

In the realm of cyclone separator design for harvesting jojoba seeds, modifying the barrel section length and inlet position while keeping the inlet dimensions constant can have a detrimental impact on performance. While the modified cyclone (III) exhibited acceptable performance, there is undoubtedly a desire to push the boundaries of this geometry model to achieve even higher cyclone efficiency. This simulation study provides a lucid overview of the most effective cyclone separator design for jojoba seed collection. As the pursuit of optimal performance persists, further modifications and optimizations will likely be explored.

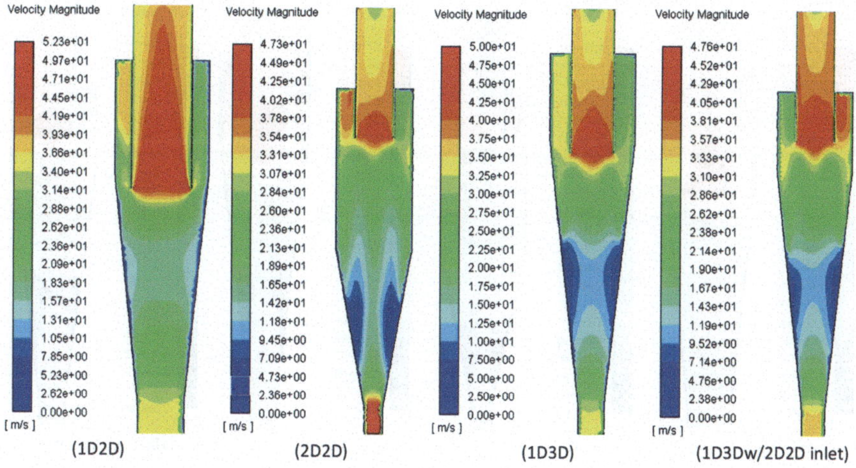

Fig. 7.22 Air velocity profiles for the standard cyclones

7.3.4 Computation Case Three (Standard Cyclones)

7.3.4.1 Fluid Profile

Cyclone separators, as already known, exhibit a remarkably detailed velocity profile, with an abundance of turbulence and intense swirls—particularly when subjected to high inlet gas velocities. It is critical to note that the simulation results obtained through this research were evaluated regarding the average velocities and velocity fluctuation levels, which play a vital role in determining the separation performance and rotational characteristics of the fluid flow. The velocity profile obtained comprehensively depicts the flow structure, commonly called the Rankine vortex type [110–112], which is prevalent in reverse flow cyclone separators. This unique structure comprises an outer free vortex and a forced vortex at the center. The flow near the walls of the cyclone is oriented toward pushing heavier particulate matter toward the outlet. As the flow reaches the conic section, its direction is reversed in the central region, moving upwards through the vortex finder, ultimately finding its way to the cyclone outflow. Precisely predicting the velocity fluctuation levels is of utmost importance for obtaining accurate separation performance results and illustrating the fluid flow's rotational properties.

As the CFD simulations commence, valuable insights into the standard cyclones' performance are gleaned. By examining the average velocity profiles presented in Fig. 7.22 and coupling them with the cutting-edge DEM software, a clear trend emerges: the highest velocity values are found within the inner flow structures at the start of the vortex finder, while the lowest is observed in the conical section. However, in the lower region of the conical part (specifically the seed's outlet), the air velocity in all standard cyclones remains well above the terminal velocity of the seeds, particularly in the 2D2D cyclone. This is due to the small diameter of the

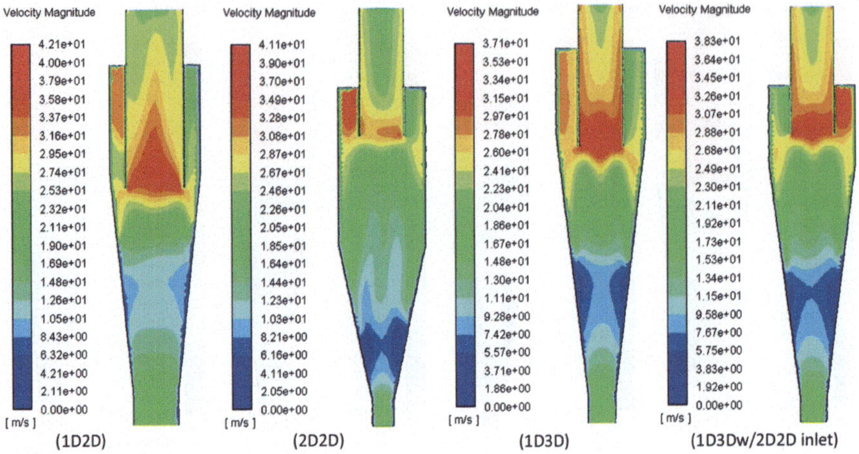

Fig. 7.23 Optimization of air velocity profiles of the standard cyclones

seed's outlet compared to other geometries, which effectively suspends the seeds within the conical section and prevents them from escaping the cyclone. Conversely, the designed cyclone exhibits a velocity profile that is optimal for catching seeds in the bottom section of the cyclone. The experimental design features a handle valve as the seed's outlet, providing the ability to adjust the gate vent and thereby regulate the velocity, ensuring the successful collection of the seeds. In summary, these findings serve as a valuable indicator of the standard cyclones' limitations and the potential benefits of innovative design in optimizing seed collection. By leveraging advanced simulation techniques and experimental methodologies, it can unlock new insights into the complex dynamics of cyclone performance, ultimately driving greater efficiency and productivity.

Upon thorough observation, it became apparent that a significant change was necessary to optimize the seed collection process in standard cyclones. To achieve this, the seed's outlet boundary conditions were modified to allow for efficient gathering at the outlet portion. This involved switching from a pressure-in boundary condition to a velocity-in boundary condition of 15 m/s—a value lower than the seed's terminal velocity [59]. In practical terms, this velocity can be regulated by adding a valve to the seed's outlet for flow control. Figure 7.23 illustrates the resulting air velocity profiles in the standard cyclones after implementing these modifications. Notably, the changes in the average velocity distribution in the vortex finder and conical section can potentially impact the separation of particles and their trajectory behavior. The effects of this modification on the behavior of particulate matter will be discussed in greater detail in subsequent sections. By implementing these changes, the collection and gathering of seeds can be made more efficient and effective, leading to improved overall performance and tremendous success in seed-related endeavors.

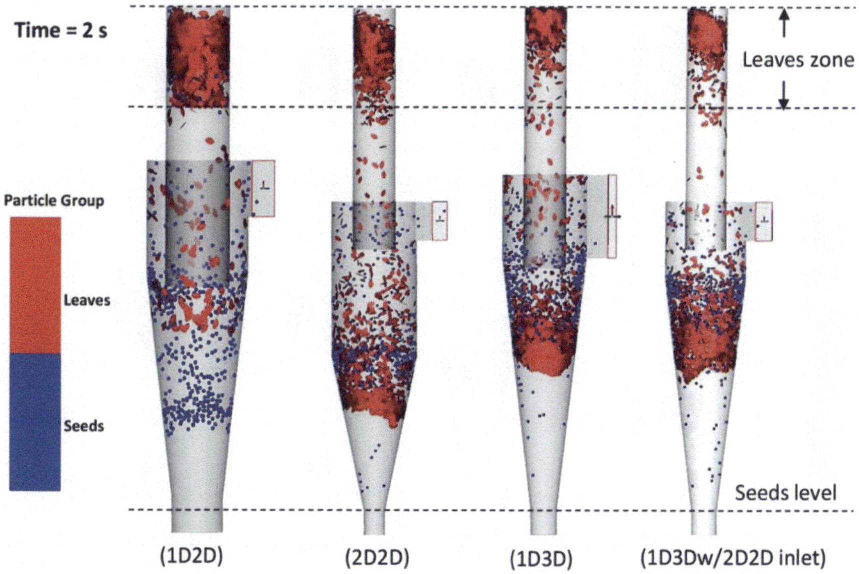

Fig. 7.24 Realizations of the particle's trajectories at 2 s of simulation time

7.3.4.2 Particle Trajectory

Separating particulate matter can be likened to a competition between centrifugal forces that draw seeds toward the cyclone wall and turbulent dispersion that scatters leaves throughout the cyclone volume in hopes of capturing them in the stream exits through the vortex finder. The geometry of the cyclone separator and the velocity profile distribution play a crucial role in determining the movement and trajectory of particles, ultimately influencing the system's performance [113, 118, 119]. Under normal boundary conditions, with the inlet pressure of seeds set, a snapshot of the particulate matter's trajectory and behavior inside a standard cyclone at different stages of the simulation process can be observed in Fig. 7.24. Starting from the second 2 of the simulation time, when the feeding of particles ceased, the 1D2D cyclone demonstrated excellent performance in separating leaves in a relatively short time, such that most of the leaves had already escaped through the vortex finder. The particles were observed to have a normal distribution in the cyclone domain, heavily influenced by the velocity profile distribution. As such, most of the particles accumulated in the conical section, awaiting the inner vortex to change direction and move upwards toward the vortex finder, encouraging the separation of leaves from the stream and pushing the seeds downwards toward the outlet. The cyclone separator's optimal design and operation depend on various factors such as geometry, velocity profile distribution, and boundary conditions. It can improve the system's performance and efficiently separate particulate matter by carefully considering these factors.

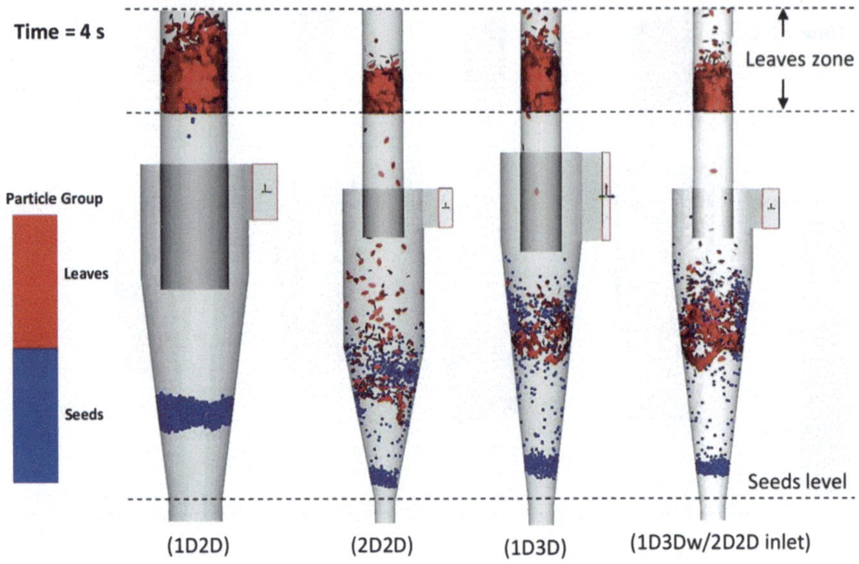

Fig. 7.25 Realizations of the particle's trajectories at 4 s of simulation time

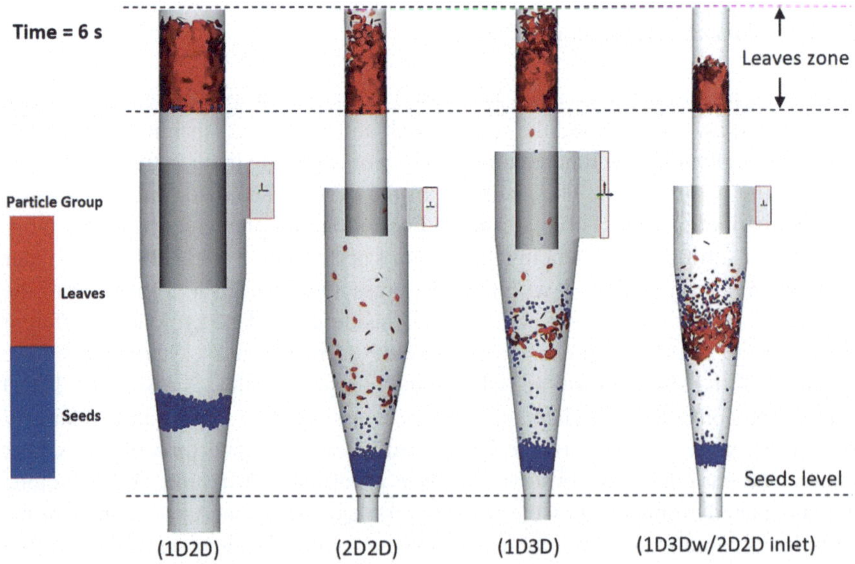

Fig. 7.26 Realizations of the particle's trajectories at 6 s of simulation time

As the simulation time increases, a noticeable trend emerges in Fig. 7.25 where standard cyclones still contain collected seeds within their domains, above the outlet level of the seeds. While the 1D2D cyclone displays an influential trend in eliminating leaves from the domain, a few seeds are lost and escape through the vortex

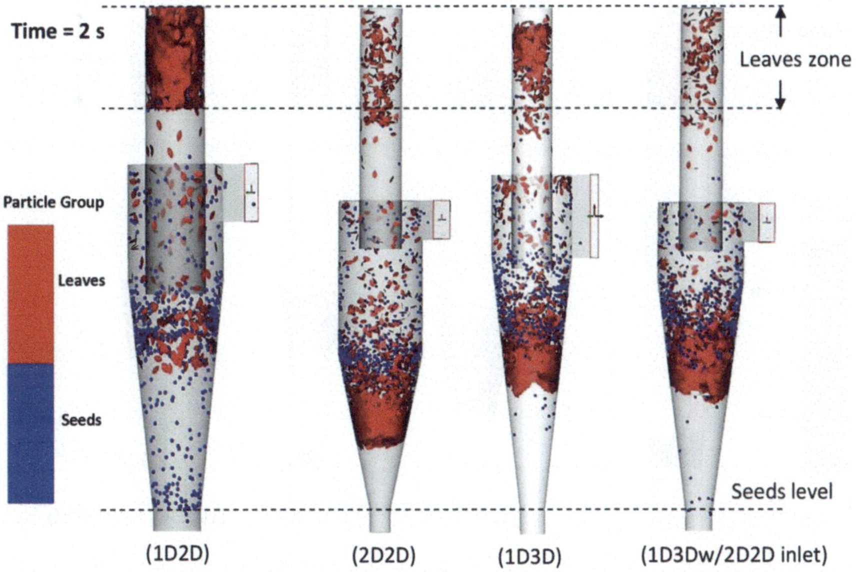

Fig. 7.27 Realizations of the optimized particle's trajectories at 2 s of simulation time

finder. Other cyclones, however, require more simulating time to gather particles in the conic part before reaching separation. At the end of the 6-s simulation, as depicted in Fig. 7.26, the number of remaining leaves in the cyclone's domains and the number of collected seeds at the outlet must be considered. These estimations significantly impact cyclones' performance, which will be discussed in the next section. Notably, the seeds in standard cyclones are typically found in the conic section and do not reach the bottom. This is primarily due to the high-velocity profiles at the bottom of the conic part, exceeding 35 m/s. This velocity is considerably higher than the seed's terminal velocity of 20 m/s and the bottom profile of the designed cyclone, as noted in the literature [59].

The 1D3D cyclone with a 2D2D inlet design exhibits a significant and powerful vortex of particles that negatively impacts the separation process compared to conventional 1D3D and 2D2D cyclones. Even though the narrow width of the conventional 1D3D inlet design causes some choking when introducing granular flow compared to the 1D3Dw/2D2D inlet design, it results in good particle separation [120, 121]. This highlights the significant impact of the inlet dimensions on cyclone performance. In addition, the outlet boundary condition for the seeds has been modified to allow them to rest at the outlet level. Comparing the optimized standard cyclone simulations at the second 2 of the simulation time, as shown in Fig. 7.27, a simple realization during DEM simulations can be predicted to estimate the performance of the standard cyclone. A significant number of seeds in the 1D3Dw/2D2D inlet and 1D2D cyclone designs have reached their macroscopically steady state at the bottom region, but the number is much higher in the 1D2D design. Additionally,

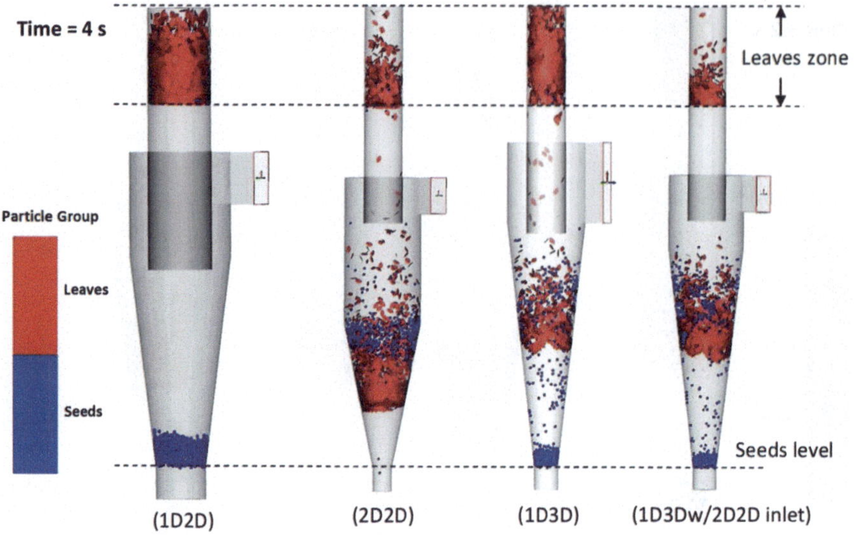

Fig. 7.28 Realizations of the optimized particle's trajectories at 4 s of the simulation time

the number of separated leaves in the 1D2D cyclone is considered to be higher than in other cyclones. Meanwhile, the 2D2D and the conventional 1D3D cyclone designs exhibit a normal distribution of seeds and leaves throughout the domain structure.

At the 4-s simulation time depicted in Fig. 7.28, the 1D2D cyclone displayed optimal separation performance, with all leaves having been separated and exited the cyclone and all seeds having accumulated at the bottom of the conic section. In contrast, the 1D3D and 1D3Dw/2D2D inlet cyclones exhibited a normal distribution of particles, with some seeds accumulating at the seed outlet. However, a significant number of leaves remained unseparated and dispersed with the seeds within the domain. The 2D2D cyclone demonstrated poor particle behavior, with most particles accumulating in the conic section and only a negligible number of seeds being separated.

In Fig. 7.29, it can be observed that cyclones 1D3D and 1D3Dw/2D2D inlet design require more time to complete particle separation. Consequently, a considerable amount of seeds and leaves are already mixed and dispersed in the domains. On the other hand, the 2D2D cyclone design shows that the particle trend accumulates in the conic section, which is believed to be caused by the chaotic velocity distribution in this area. This behavior may affect the particle interaction forces and hinder seeds from falling downward. Therefore, the 1D2D cyclone design appears to have the best separation efficiency since no particles were present inside the domain, indicating superior particle separation performance.

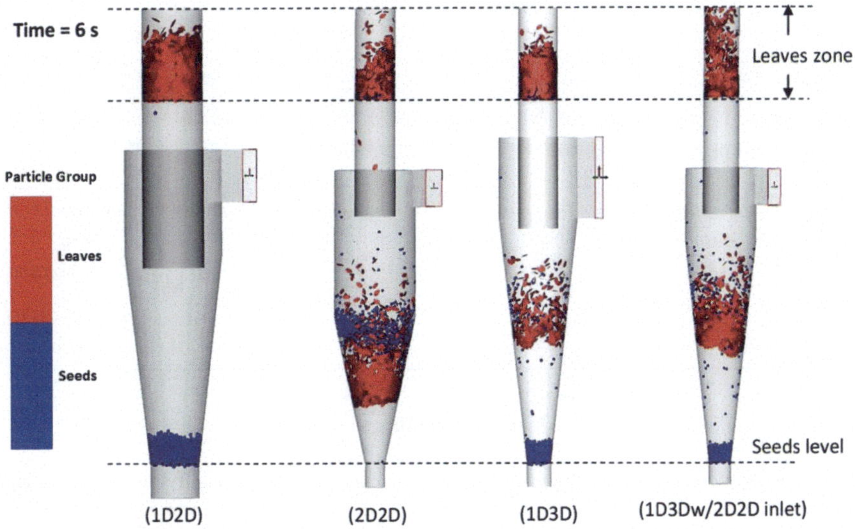

Fig. 7.29 Realizations of the optimized particle's trajectories at 6 s of the simulation time

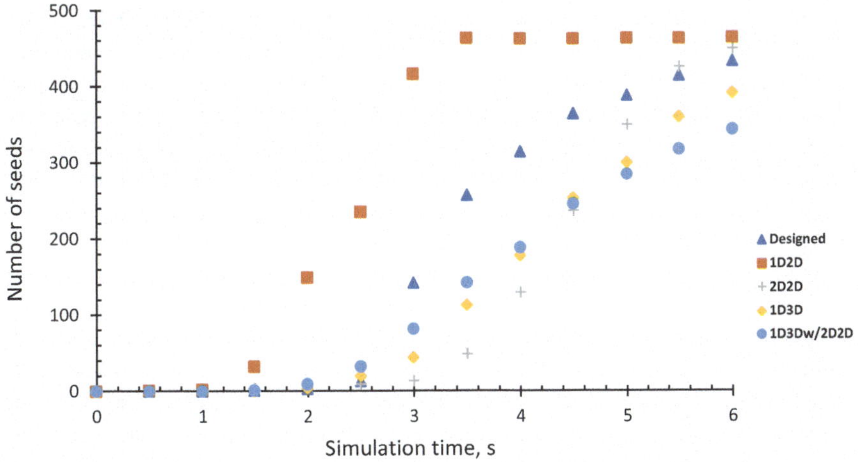

Fig. 7.30 Number of separated seeds during the simulation time in the ordinary case

7.3.4.3 Performance of Standard Cyclones

In seed separation technology, simulation plays a crucial role in determining the performance of various cyclones. In a recent study, the ordinary and modified cases of 1D2D, 2D2D, 1D3D, and 1D3Dw/2D2D cyclones were investigated, with Figs. 7.30 and 7.31 showcasing the number of separated and collected seeds,

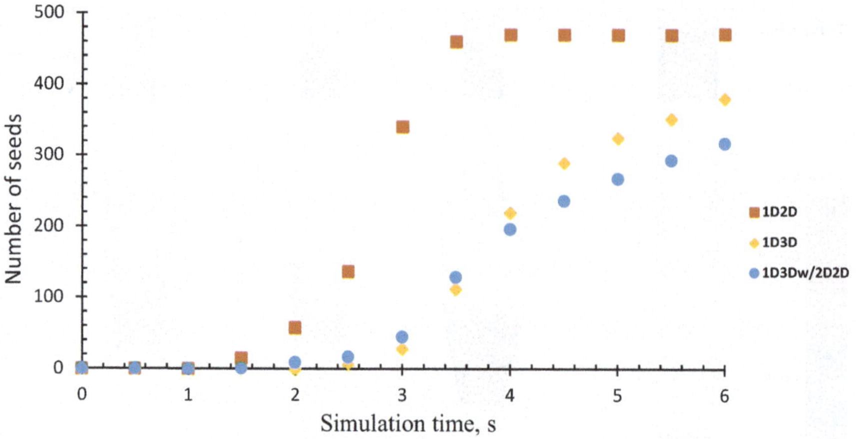

Fig. 7.31 Number of separated seeds during the simulation after modifying the operating conditions

respectively. Upon analysis, it was found that the total number of separated seeds in the ordinary simulation was 464, 449, 392, and 343 for the 1D2D, 2D2D, 1D3D, and 1D3Dw/2D2D cyclones, respectively. However, upon modifying the operating conditions, the number of collected seeds in the 1D2D cyclone increased by a modest 1.7%. On the other hand, the 1D3D and 1D3Dw/2D2D cyclones decreased the number of seeds collected by 2.5% and 7.3%, respectively. The 2D2D cyclone performed the worst in terms of optimization, with no separated seeds. In terms of time, the curve tendency of the 1D2D cyclone reached a stable state at the simulation time of 3.5 s, where the seeds were successfully separated. In contrast, the other cyclones maintained a proportional relationship with time, requiring more simulation time and intensive power to complete the seed separation process. Furthermore, a comparison between the designed and optimized 1D2D cyclones revealed that the latter could collect more seeds by 8.2%, albeit with 28 seed particles escaping through the vortex finder. The study also found that the cyclone vortex finder dimensions were crucial in determining cyclone performance. Its effects were more pronounced than changing other cyclone dimensions, as observed in the literature [122, 123].

On the other hand, the number of leaves that escaped through the vortex finder represents the seed separation quality investigated. Figures 7.32 and 7.33 compare the separated leaves numbers in the ordinary and modified cases in conjunction with the time step. It was observed that the curve tendency of the 1D2D cyclone has a constant function that starts approximately 2.2 s into the simulation time, which means all of the leaves had already escaped from the domain. According to the designed cyclone result, about 1479 leaves particle escaped. Nevertheless, many leaves are already separated but still inside the cyclone at the top portion of the barrel because the inlet portion design is fixed by 70 mm below the top surface of the cyclone.

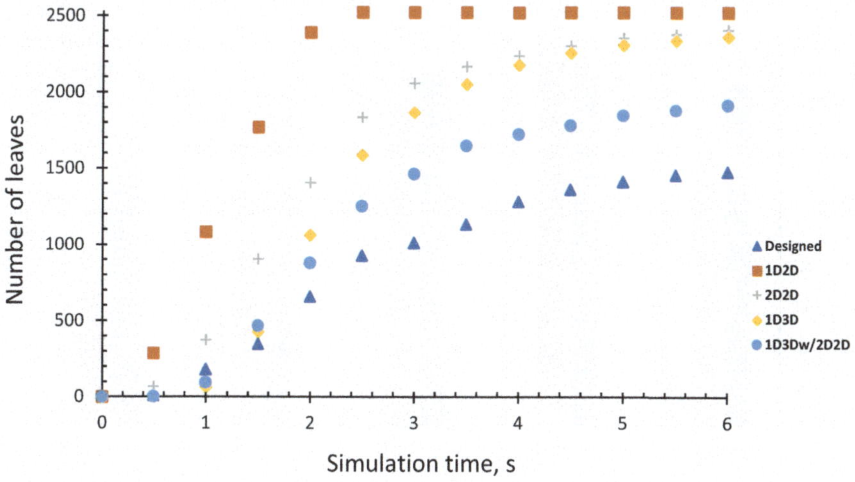

Fig. 7.32 Number of separated leaves during the simulation time in the ordinary case

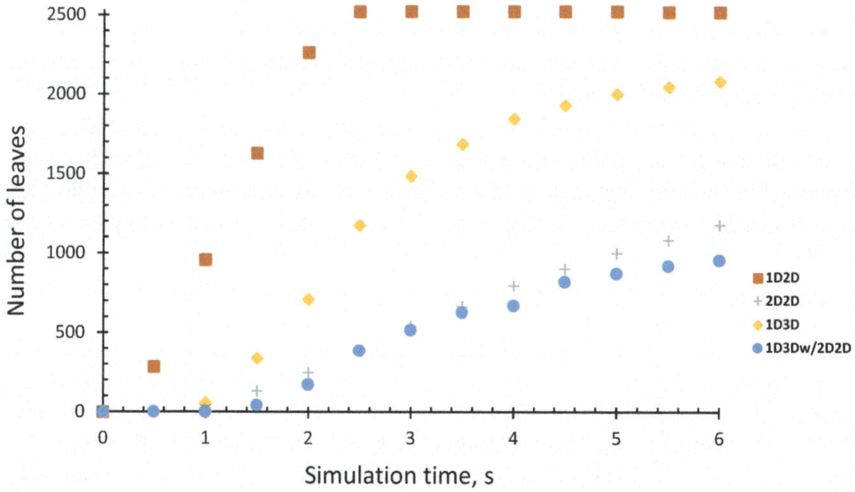

Fig. 7.33 Number of separated leaves during the simulation after modifying the operating conditions

From the expected simulation result, it can be derived that the number of separated leaves reached 2406, 2367, and 1916 in the 2D2D cyclone designs, followed by 1D3D and 1D3Dw/2D2D cyclones, respectively. After modifying the operating condition, the 1D2D cyclone curve still has the same tendency, while the number of separated seeds in the 2D2D, 1D3D, 1D3Dw/2D2D cyclones are severely decreased by 51%, 12%, and 50%, respectively. So, under these obtained data, the 1D2D

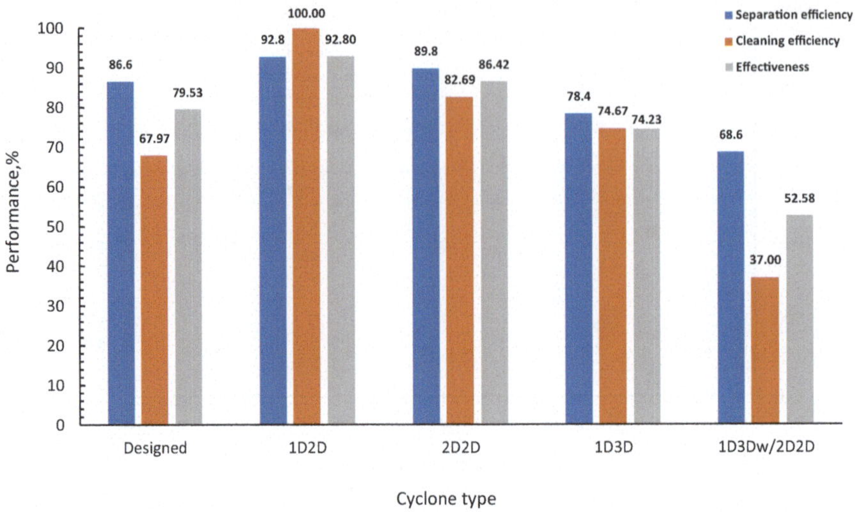

Fig. 7.34 Performance comparison of the investigated cyclones under ordinary simulation

cyclone design shows the best and most high-performance trend to deal with the investigated particulate matter.

The performance of a cyclone can be calculated and analyzed based on the results obtained from DEM simulations and Eqs. (7.1), (7.2), (7.3), and (7.4) as discussed in the previous section. By analyzing the number of separated particles, one can quickly determine the efficiency of the cyclone. It is worth noting that in the standard cyclones simulation Fig. 7.34, the separating efficiency of the seeds is not a realistic representation. This is because the separated seed particles do not collect at the bottom of the conic part, but this result is still helpful in evaluating the cyclonic process. Based on the analysis, the 1D2D cyclone has the most impressive performance trend, with separating and cleaning efficiencies reaching 92.8% and 100%, respectively. This high level of performance makes the cyclone very effective. However, it is unfortunate that the designed cyclone performs the worst in cleaning efficiency and effectiveness, recording only 29.8% and 51.2%, respectively. Nevertheless, the separating efficiency is still considered reasonable, at 86.6%. The decline in the performance of this cyclone can be attributed to the fact that only the number of leaves that escaped from the vortex finder was considered during the calculation. Despite this, the designed cyclone is still the best option for separating jojoba seeds from leaves, allowing them to rest at the bottom of the conic part. This allows for identifying the most effective cyclone for a given operation, thereby improving overall efficiency and productivity.

The performance of these cyclones was re-evaluated and compared by optimizing the operating conditions to allow the collection of seed particles at the bottom level of standard cyclones. The results, illustrated in Fig. 7.35, showed that the 1D2D cyclone outperformed the other cyclones, with an increase of 1.7% in

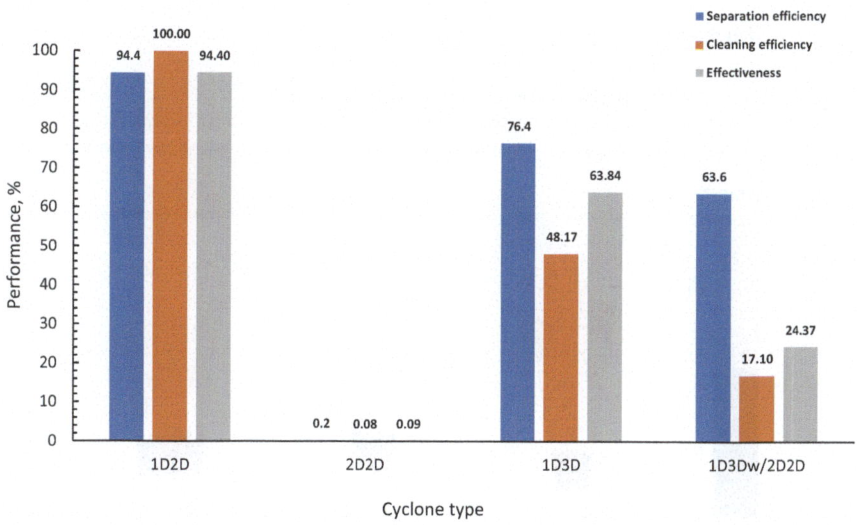

Fig. 7.35 Performance comparison of the investigated cyclones under modifying the operation condition

separation efficiency and effectiveness. However, this improvement was accompanied by a 5.6% loss of seed particles that escaped with the leaves. On the other hand, the 1D3D and the 1D3Dw/2D2D inlet decreased by 2.5% and 7.3% in separation efficiency, respectively, resulting in a significant drop in cleaning efficiency by 53.8% and 35.6%. These deteriorated performance parameters had a negative impact on the overall effectiveness of these cyclones. Interestingly, the 2D2D cyclone exhibited the worst performance behavior, with the separating and cleaning efficiencies almost equal to zero. This highlights the importance of optimizing operating conditions and carefully selecting cyclone configurations to achieve optimal performance and avoid losses in the separation process.

7.3.5 Comparison of Performance of Various Cyclones

This section compares the designed and optimized cyclones in the previous three cases, specifically their efficiency in separating jojoba seeds from leaves at a high particle loading ratio of 0.5. The separation efficiency, which represents the number of collected seeds at the bottom of the cyclone, is considered the most critical parameter in this comparison due to the high value of jojoba seeds. Upon the completion of the simulation time, as illustrated in Figs. 7.36 and 7.37, the designed cyclone exhibits that most of the seed particles have been separated from the leaves. However, some seeds remain gathered in the conic part and require more simulation

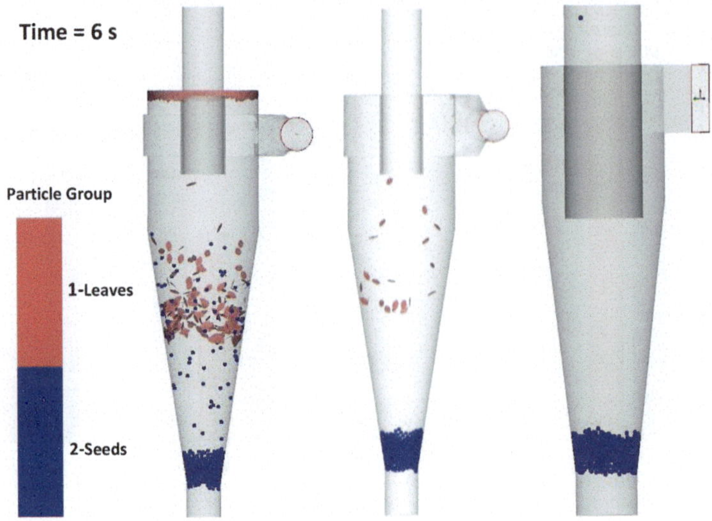

Fig. 7.36 Particle behavior comparison between the designed and optimized cyclones; (**a**) designed, (**b**) cyclone III, (**c**) 1D2D cyclone

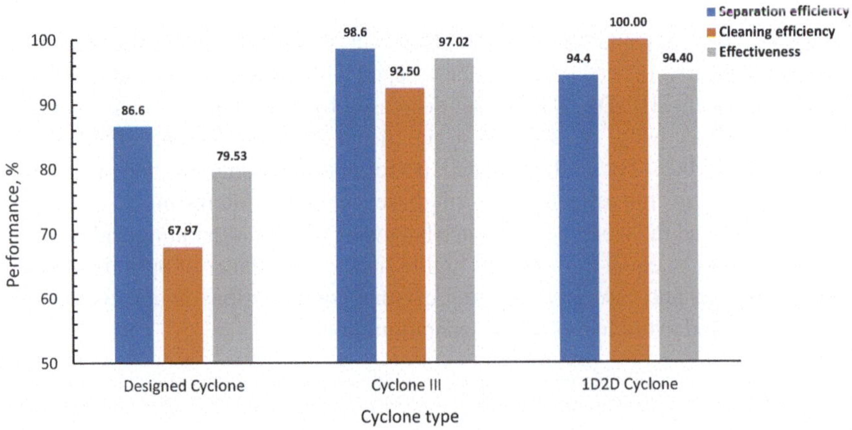

Fig. 7.37 Performance comparison between the designed and the optimized cyclones

time, resulting in considerable power consumption to achieve the desired separation. Furthermore, many leaves were already separated but remained inside the cyclone at the top of the cylinder part. This behavior may be attributed to the fixed inlet portion design, which is situated 70 mm below the top surface of the cyclone. As a result, the performance of the designed cyclone is suboptimal for this separation process using large particles, with a separation efficiency of only 86.6%. Therefore, optimizing the design and increasing the performance of the designed cyclone is essential. After optimization, the 1D2D cyclone demonstrates good particle trajectory behavior, with a cleaning efficiency of 100%. However, cyclone (III) still boasts the best performance. The separation efficiency of the cyclone (III) is 4.2% higher than that of the 1D2D cyclone, indicating a higher number of collected seeds. Consequently, the number of seed losses that remain in the cyclone with the leaves is much higher for the 1D2D cyclone than cyclone (III). Additionally, the effectiveness of cyclone (III) is 2.62% higher than the 1D2D cyclone. This suggests that changing the cyclone inlet geometry has a more significant impact on performance than changing the entire geometry. Therefore, cyclone (III) is deemed the best design for this investigation to separate jojoba seeds from their leaves.

7.4 Summary

The strength of the CFD–DEM approach has been harnessed to simulate and optimize cyclonic processes using macroscopic granular systems. A soft-sphere approach within DEM was utilized to develop these multiphase flow simulation processes. Soft sphere models offer the advantage of representing multiple particle contacts, which is crucial for modeling bulk grain systems. The particle models varied based on the type of grain. This research marks a pioneering approach in cyclonic separation and purification, employing actual shape modeling of particles. The CFD–DEM coupling approach was applied to predict the behavior of macroscopic particles, a technique increasingly employed in agricultural systems for low-power simulations. However, while this coupling has been found to help study chemical particle processes, its application to modeling actual macroscopic agricultural grains is novel.

The comparison between simulation parameters and experimental results for the designed cyclone demonstrates the predictiveness of the CFD–DEM model. This model can be extended to various aerodynamic processes relevant to the agricultural field. The separation efficiency aligns closely with experimental data, deviating only by 1.26%. Similarly, the cleaning efficiency exhibits a small gap of 1.94% less than the experimental results. Consequently, cyclone effectiveness diminishes by 6.3%, a value reliant on both separation and cleaning efficiencies. Four modifications to cyclone geometry were made by adjusting inlet dimensions and positions to optimize cyclone performance. Analysis of simulation data reveals that the third modification, cyclone (III), featuring double the original inlet size, offers the best predictive fit for high cyclone performance, boasting 98.6% separation efficiency, 92.5% cleaning efficiency, and 97% cyclone effectiveness. Notably, altering inlet dimensions beyond those in cyclone (III) negatively impacts cyclone performance. Moreover, reducing barrel length and changing the original inlet position on the cyclone barrel section result in poor cyclone performance.

Additionally, the performance of the designed cyclone was compared with four standard cyclones, namely, 1D2D, 2D2D, 1D3D, and 1D3D, with a 2D2D inlet, to determine optimal performance. Initial simulation results with ordinary boundary conditions were inconclusive, as seeds did not settle at the cyclones' bottoms. This prompted the urgent need to modify the cyclones' operating conditions. The 1D2D cyclone exhibited the most promising performance trend, with 100% cleaning efficiency following modifications. However, the percentage of seed loss escaping with the leaves increased to 5.6%.

A comparison of the modified cyclones with optimized standard cyclones revealed that cyclone (III) boasts a 4.2% higher separation efficiency than the 1D2D cyclone, resulting in more collected seeds. Additionally, the percentage of seed loss accompanying leaves in cyclone (III) is considerably lower than in the 1D2D cyclone. Furthermore, cyclone (III) exhibited 2.62% higher effectiveness than the 1D2D cyclone, indicating that altering cyclone inlet geometry proves more impactful than changing the entire geometry. Hence, cyclone (III) emerges as the optimal design for this investigation, specifically for separating jojoba seeds from their leaves.

Based on these computational findings, cyclone (III) is recommended as the superior choice for harvesting jojoba seeds from soil surfaces and separating them from impurities. This coupling technique and simulation methodology offer avenues for enhancing the aerodynamic systems of agricultural machines without substantial cost or effort. The characteristics gleaned from this investigation lay the foundation for developing and optimizing the structural and technological parameters of aerodynamic systems in agricultural machinery.

References

1. Spasov GH, Rossi R, Vanossi A et al (2024) A critical analysis of the CFD-DEM simulation of pharmaceutical aerosols deposition in extra-thoracic airways. Comput Biol Med 629:107948. https://doi.org/10.1016/j.ijpharm.2022.122331
2. Spasov GH, Rossi R, Vanossi A et al (2022) A critical analysis of the CFD-DEM simulation of pharmaceutical aerosols deposition in extra-thoracic airways. Int J Pharm 629:122331. https://doi.org/10.1016/j.ijpharm.2022.122331
3. Ariane M, Sommerfeld M, Alexiadis A (2018) Wall collision and drug-carrier detachment in dry powder inhalers: using DEM to devise a sub-scale model for CFD calculations. Powder Technol 334:65–75. https://doi.org/10.1016/j.powtec.2018.04.051
4. Tong ZB, Yang RY, Yu AB (2017) CFD-DEM study of the aerosolisation mechanism of carrier-based formulations with high drug loadings. Powder Technol 314:620–626. https://doi.org/10.1016/j.powtec.2016.10.004
5. Zhu Q, Kakhi M, Jayasundara C et al (2023) CFD-DEM investigation of the effects of aperture size for a capsule-based dry powder inhaler. Int J Pharm 647:123556. https://doi.org/10.1016/j.ijpharm.2023.123556
6. Yazdanfar Z, Lester D, Robert D, Setunge S (2021) A novel CFD-DEM upscaling method for prediction of scour under live-bed conditions. Ocean Eng 220:108442. https://doi.org/10.1016/j.oceaneng.2020.108442
7. Di Y, Zhao L, Mao J (2021) A resolved CFD-DEM method based on the IBM for sedimentation of dense fluid-particle flows. Comput Fluids 226:104968. https://doi.org/10.1016/j.compfluid.2021.104968

8. Sun R, Xiao H, Sun H (2017) Realistic representation of grain shapes in CFD–DEM simulations of sediment transport with a bonded-sphere approach. Adv Water Resour 107:421–438. https://doi.org/10.1016/j.advwatres.2017.04.015
9. Ma H, Li B, Zhang S (2023) Erosion mechanism for submarine pipeline on graded sediments using a coupled PORO-CFD-DEM model. Comput Geotech 162:105672. https://doi.org/10.1016/j.compgeo.2023.105672
10. Sun R, Xiao H (2016) SediFoam: a general-purpose, open-source CFD-DEM solver for particle-laden flow with emphasis on sediment transport. Comput Geosci 89:207–219. https://doi.org/10.1016/j.cageo.2016.01.011
11. Parsi M, Najmi K, Naja F et al (2014) A comprehensive review of solid particle erosion modeling for oil and gas wells and pipelines applications. J Nat Gas Sci Eng 21:850–873. https://doi.org/10.1016/j.jngse.2014.10.001
12. Abdulla A (2011) Estimating erosion in oil and gas pipe line due to sand presence. Blekinge Institute of Technology, Karlskrona
13. Bahrami B, Mohsenpour S, Shamshiri Noghabi HR et al (2019) Estimation of flow rates of individual phases in an oil-gas-water multiphase flow system using neural network approach and pressure signal analysis. Flow Meas Instrum 66:28–36. https://doi.org/10.1016/j.flowmeasinst.2019.01.018
14. Wang X, Gong L, Li Y, Yao J (2023) Developments and applications of the CFD-DEM method in particle–fluid numerical simulation in petroleum engineering: a review. Appl Therm Eng 222:119865. https://doi.org/10.1016/j.applthermaleng.2022.119865
15. Wang S, Wu K, Chu S et al (2023) Comprehensive DEM-CFD model and thermal uniformity optimization for biomass catalytic pyrolysis reactor. Energy Convers Manag 291:117321. https://doi.org/10.1016/j.enconman.2023.117321
16. Porcaro C, Saeedipour M (2023) Hemolysis prediction in bio-microfluidic applications using resolved CFD-DEM simulations. Comput Methods Prog Biomed 231:107400. https://doi.org/10.1016/j.cmpb.2023.107400
17. Zhao J, Haghnegahdar A, Feng Y et al (2022) Prediction of the carrier shape effect on particle transport, interaction and deposition in two dry powder inhalers and a mouth-to-G13 human respiratory system: a CFD-DEM study. J Aerosol Sci 160:105899. https://doi.org/10.1016/j.jaerosci.2021.105899
18. Djukic T, Topalovic M, Filipovic N (2022) Validation of lattice Boltzmann based software for blood flow simulations in complex patient-specific arteries against traditional CFD methods. Math Comput Simul 203:957–976. https://doi.org/10.1016/j.matcom.2022.07.027
19. Aleksić I, Jelena Parojčić ZĐ (2024) Computational fluid dynamics: applications in pharmaceutical technology. In: Djuris J (ed) Computer-aided applications in pharmaceutical technology, Woodhead publishing series in biomedicine, 2nd edn. Woodhead Publishing, Cambridge, pp 285–315
20. Niu Q, Wang NX (2019) Study of heat transfer by using DEM–CFD method in a randomly packed pebble-bed reactor. Nucl Sci Tech 30:1–12. https://doi.org/10.1007/s41365-019-0556-8
21. Ding W, Chen R, Tian W et al (2023) Numerical investigation of dynamic characteristics of debris bed formation based on CFD-DEM method. Ann Nucl Energy 180. https://doi.org/10.1016/j.anucene.2022.109492
22. Mardus-Hall R, Ho M, Pastrello A, Yeoh G (2020) 3-way coupled thermohydraulic-discrete element-neutronic simulation of solid fuel, molten salt reactor. Ann Nucl Energy 135:106973. https://doi.org/10.1016/j.anucene.2019.106973
23. Horabik J, Molenda M (2016) Parameters and contact models for DEM simulations of agricultural granular materials: a review. Biosyst Eng 147:206–225. https://doi.org/10.1016/j.biosystemseng.2016.02.017
24. Zhengdao L, Qingjie W, Hongwen L et al (2019) Fertilizer injecting route analysis and test for air-blowing seed-fertilizer hole-applicator via CFD-DEM coupling. Trans Chin Soc Agric Eng 35:18–25. https://doi.org/10.11975/j.issn.1002-6819.2019.21.003

25. Xi X, Wang R, Wang X et al (2023) Parametric optimization and experimental verification of multi-fertilizer mixing by air blowing and blade stirring based on discrete element simulations. Comput Electron Agric 210:107895. https://doi.org/10.1016/j.compag.2023.107895
26. Liu R, Liu Z, Zhao J et al (2022) Optimization and experiment of a disturbance-assisted seed filling high-speed vacuum seed-metering device based on DEM-CFD. Agriculture 12. https://doi.org/10.3390/agriculture12091304
27. Song X, Dai F, Zhang F et al (2023) Calibration of DEM models for fertilizer particles based on numerical simulations and granular experiments. Comput Electron Agric 204:107507. https://doi.org/10.1016/j.compag.2022.107507
28. Guzman L, Chen Y, Landry H (2023) Coupled CFD-DEM simulation of seed flow in horizontal-vertical tube transition. Processes 11. https://doi.org/10.3390/pr11030909
29. Kim YS, Lee SD, Baek SM et al (2022) Development of DEM-MBD coupling model for draft force prediction of agricultural tractor with plowing depth. Comput Electron Agric 202:107405. https://doi.org/10.1016/j.compag.2022.107405
30. Liu Y, Li Y, Dong Y et al (2022) Development of a variable-diameter threshing drum for rice combine harvester using MBD – DEM coupling simulation. Comput Electron Agric 196:106859. https://doi.org/10.1016/j.compag.2022.106859
31. Saunders C, Ucgul M, Godwin RJ (2021) Discrete element method (DEM) simulation to improve performance of a mouldboard skimmer. Soil Tillage Res 205:104764. https://doi.org/10.1016/j.still.2020.104764
32. Tang H, Xu C, Zhao J, Wang J (2023) Formation and steady state characteristics of flow field effect in the header of a stripping prior to cutting combine harvester with CFD. Comput Electron Agric 211:107959. https://doi.org/10.1016/j.compag.2023.107959
33. Zhu D, Shi M, Yu C et al (2023) Tool-straw-paddy soil coupling model of mechanical rotary-tillage process based on DEM-FEM. Comput Electron Agric 215:108410. https://doi.org/10.1016/j.compag.2023.108410
34. Bahrami M, Naderi-Boldaji M, Ghanbarian D, Keller T (2023) Discrete element modelling of stress propagation in soil under a rigid wheel in a soil bin: a simulation of probe inducing stress deviation and wheel speed. Biosyst Eng 230:159–170. https://doi.org/10.1016/j.biosystemseng.2023.04.013
35. Tang Z, Gong H, Wu S et al (2023) Modelling of paddy soil using the CFD-DEM coupling method. Soil Tillage Res 226:105591. https://doi.org/10.1016/j.still.2022.105591
36. Qi L, Chen Y, Sadek M (2019) Simulations of soil flow properties using the discrete element method (DEM). Comput Electron Agric 157:254–260. https://doi.org/10.1016/j.compag.2018.12.052
37. Wu P, Zhang X, Zeng Z, Chen Y (2024) DEM simulation of subsoiling for soil disturbance as affected by soil layering and working speed. Smart Agric Technol 7:100385. https://doi.org/10.1016/j.atech.2023.100385
38. Yang X, Xu Z, Chai J et al (2023) Numerical investigation of the seepage mechanism and characteristics of soil-structure interface by CFD-DEM coupling method. Comput Geotech 159:105430. https://doi.org/10.1016/j.compgeo.2023.105430
39. Liu H, He Y, Tang T, Zhai M (2023) Prediction of the non-uniform drying of wet grain in a fluidised-bed based on a modified CFD-DEM drying model. Biosyst Eng 236:103–119. https://doi.org/10.1016/j.biosystemseng.2023.10.009
40. Azmir J, Hou Q, Yu A (2019) CFD-DEM simulation of drying of food grains with particle shrinkage. Powder Technol 343:792–802. https://doi.org/10.1016/j.powtec.2018.11.097
41. Azmir J, Hou Q, Yu A (2020) CFD-DEM study of the effects of food grain properties on drying and shrinkage in a fluidised bed. Powder Technol 360:33–42. https://doi.org/10.1016/j.powtec.2019.10.021
42. Duga AT, Delele MA, Ruysen K et al (2017) Development and validation of a 3D CFD model of drift and its application to air-assisted orchard sprayers. Biosyst Eng 154:62–75. https://doi.org/10.1016/j.biosystemseng.2016.10.010

43. Tang Q, Zhang R, Chen L et al (2022) Influence of the atmospheric boundary layer stability on aerial spraying studied by computational fluid dynamics. Biosyst Eng 215:170–187. https://doi.org/10.1016/j.biosystemseng.2022.01.007
44. Qiu W, Guo H, Zheng H et al (2022) CFD modelling to analyze the droplets deposition behavior on vibrating rice leaves. Comput Electron Agric 201:107330. https://doi.org/10.1016/j.compag.2022.107330
45. Salcedo R, Vallet A, Granell R et al (2017) Eulerian–Lagrangian model of the behaviour of droplets produced by an air-assisted sprayer in a citrus orchard. Biosyst Eng 154:76–91. https://doi.org/10.1016/j.biosystemseng.2016.09.001
46. Yan C, Niu C, Ma S et al (2022) CFD models as a tool to analyze the deformation behavior of grape leaves under an air-assisted sprayer. Comput Electron Agric 198:107112. https://doi.org/10.1016/j.compag.2022.107112
47. Xu Z, Mao XY, Gu YJ et al (2023) Analysis of hydrodynamic filtration performance in a cross-step filter for drip irrigation based on the CFD–DEM coupling method. Biosyst Eng 232:114–128. https://doi.org/10.1016/j.biosystemseng.2023.07.004
48. Zheng K, Bo Y, Zhang Y et al (2022) CFD simulation and experimental study of the irrigation performance of new ebb-and-flow soilless cultivation troughs. Comput Electron Agric 202. https://doi.org/10.1016/j.compag.2022.107394
49. Bouhoun Ali H, Bournet PE, Cannavo P, Chantoiseau E (2019) Using CFD to improve the irrigation strategy for growing ornamental plants inside a greenhouse. Biosyst Eng 186:130–145. https://doi.org/10.1016/j.biosystemseng.2019.06.021
50. Tang P, Li H, Issaka Z, Chen C (2017) Impact forces on the drive spoon of a large cannon irrigation sprinkler: simple theory, CFD numerical simulation and validation. Biosyst Eng 159:1–9. https://doi.org/10.1016/j.biosystemseng.2017.04.005
51. Fengwei G, Youqun Z, Feng W et al (2022) Simulation analysis and experimental validation of conveying device in uniform rrushed straw throwing and seed-sowing machines using CFD-DEM coupled approach. Comput Electron Agric 193:106720. https://doi.org/10.1016/j.compag.2022.106720
52. Wang Y, He M, Yan Q, Zheng Z (2023) DEM-CFD simulation and seed orientation evaluation of a self-suction wheat shooting device. Powder Technol 427:118746. https://doi.org/10.1016/j.powtec.2023.118746
53. Xu J, Sun S, He Z et al (2023) Design and optimisation of seed-metering plate of air-suction vegetable seed-metering device based on DEM-CFD. Biosyst Eng 230:277–300. https://doi.org/10.1016/j.biosystemseng.2023.04.018
54. Tang H, Xu C, Zhao J, Wang J (2023) Stripping mechanism and loss characteristics of a stripping-prior-to-cutting header for rice harvesting based on CFD-DEM simulations and bench experiments. Biosyst Eng 229:116–136. https://doi.org/10.1016/j.biosystemseng.2023.03.023
55. Tang H, Xu W, Zhao J et al (2023) Comparison of rice straw compression characteristics in vibration mode based on discrete element method. Biosyst Eng 230:191–204. https://doi.org/10.1016/j.biosystemseng.2023.04.009
56. Wang Y, Su W, Lai Q et al (2023) Simulation and measurement of the suction force on ellipsoidal seeds in an air-suction seed-metering device. Biosyst Eng 232:97–113. https://doi.org/10.1016/j.biosystemseng.2023.07.001
57. Zhang Z, Mei F, Xiao P et al (2023) Discrete element modelling and simulation parameters calibration for the compacted straw cube. Biosyst Eng 230:301–312. https://doi.org/10.1016/j.biosystemseng.2023.04.019
58. Hu H, Zhou Z, Wu W et al (2021) Distribution characteristics and parameter optimisation of an air-assisted centralised seed-metering device for rapeseed using a CFD-DEM coupled simulation. Biosyst Eng 208:246–259. https://doi.org/10.1016/j.biosystemseng.2021.06.005

59. El-Emam MA, Ahmed SF, Sabah MA et al (2019) Design and construction of a pneumatic harvesting and cleaning machine for jojoba seeds. Eng Agric Environ Food 12:451–459. https://doi.org/10.1016/j.eaef.2019.08.001
60. Slak MD, Prasad RO, Bakker A, Boysan F (2000) Advances in cyclone modelling using unstractured grids. Trans Inst Chem Eng 78:1098–1104. https://doi.org/10.1205/026387600528373
61. Chu KW, Wang B, Xu DL et al (2011) CFD-DEM simulation of the gas-solid flow in a cyclone separator. Chem Eng Sci 66:834–847. https://doi.org/10.1016/j.ces.2010.11.026
62. Lingjuan W, Parnell CB, Shaw BW (2002) A study of the cyclone fractional efficiency curves. Agric Eng Int CIGR J Sci Res Dev IV:1–14
63. Surmen A, Avci A, Karamangil MI (2011) Prediction of the maximum-efficiency cyclone length for a cyclone with a tangential entry. Powder Technol 207:1–8. https://doi.org/10.1016/j.powtec.2010.10.002
64. Yoshida H, Kwan-Sik Y, Fukui K et al (2003) Effect of apex cone height on particle classification performance of a cyclone separator. Adv Powder Technol 14:263–278. https://doi.org/10.1163/15685520360685938
65. Elsayed K, Lacor C (2010) Optimization of the cyclone separator geometry for minimum pressure drop using mathematical models and CFD simulations. Chem Eng Sci 65:6084–6058. https://doi.org/10.1016/j.powtec.2014.09.038
66. Taiwo MI, Namadi MA, Mokwa B (2016) Design and analysis of cyclone dust separator. Am J Eng Res 5:130–134
67. Simonyan KJ, Yiljep YD (2008) Investigating grain separation and cleaning efficiency distribution of a conventional stationary rasp-bar sorghum thresher. Agric Eng Int CIGR J X:1–13
68. Brooks WH (1978) Jojoba—a North American desert shrub; its ecology, possible commercialization, and potential as an introduction into other arid regions. J Arid Environ 1:227–236. https://doi.org/10.1016/S0140-1963(18)31725-7
69. Ismail AMA (1988) The ecological and agronomic role of seed polymorphism in Simmondsia chinensis. J Arid Environ 14:35–42. https://doi.org/10.1016/S0140-1963(18)31094-2
70. Sardana J, Batra A (1998) In vitro regeneration of jojoba (Simmondsia chinensis): a plant of high potential. Adv Plant Sci 11:143–146
71. Castellanos AE, Molina FE (1990) Differential survivorship and establishment in Simmondsia chinensis (jojoba). J Arid Environ 19:65–76. https://doi.org/10.1016/S0140-1963(18)30830-9
72. Coates W, Yazici E (1990) Aerodynamic characteristics of jojoba seeds and like-sized stones. Trans ASAE 33:1449–1456. https://doi.org/10.13031/2013.31492
73. Mayer AM, Poljakoff-Mayber A (1982) The germination of seeds, 3rd edn. Pergamon, Burlington
74. Kant K, Sharma B, Tyagi M (1983) Effects of maturation environment on seed size and subsequent plant growth in peas (Pisum sativum). Exp Agric 19:333–336. https://doi.org/10.1017/S0014479700012783
75. Khan D, Zaki MJ, Rauf A (2015) Some observations on seeds and seedlings of jojoba [Simmondsia Chinensis (link) C. K. Schneider (family: Simmondsiaceae)]—a native of Sonoran Desert—cultivated in Pakistan. Int J Biol Biotechnol 12:649–666
76. Wang YJ, Chung DS, Spillman CK et al (1994) Evaluation of laboratory grain cleaning and separating equipment – Part I. Trans ASAE 37:507–513
77. Sokhansanj S, Fang R, Opoku A (1999) Influence of physical properties and operating conditions on particle segregation on gravity table. Appl Eng Agric 15:495–499. https://doi.org/10.13031/2013.5809
78. Pouliot Y, Conway V, Leclerc P (2014) Separation and concentration technologies in food processing. In: Clark S, Jung S, Lamsal B (eds) Food processing: principles and applications, second. Wiley, Québec, pp 33–60
79. Elghobashi S (1994) On predicting particle-laden turbulent flows. Appl Sci Res 52:309–329. https://doi.org/10.1007/BF00936835
80. Elghobashi S (1991) Particle-laden turbulent flows: direct simulation and closure models. Appl Sci Res 48:301–314. https://doi.org/10.1007/BF02008202

81. ESSS-Rocky (2018) Rocky-Dem Technical Manual. In: Models used in Rocky. ESSS, Rio de Janeiro, Brazil
82. Zhao H, Zhao Y (2019) CFD–DEM simulation of pneumatic conveying in a horizontal channel. Int J Multiph Flow 118:64–74. https://doi.org/10.1016/j.ijmultiphaseflow.2019.06.003
83. Potapov A, Chen X, Donohue T, Wheeler C (2013) Computer simulation of airflow around transfer chutes via linked discrete element method – computational fluid dynamics approach. In: 11th international congress on bulk materials storage, handling and transportation
84. Gimbun J, Chuah TG, Fakhru'l-Razi A, Choong TSY (2005) The influence of temperature and inlet velocity on cyclone pressure drop: a CFD study. Chem Eng Process 44:7–12. https://doi.org/10.1016/j.cep.2004.03.005
85. Wan G, Sun G, Xue X, Shi M (2008) Solids concentration simulation of different size particles in a cyclone separator. Powder Technol 183:94–104. https://doi.org/10.1016/j.powtec.2007.11.019
86. Xiang RB, Lee KW (2005) Numerical study of flow field in cyclones of different height. Chem Eng Process 44:877–883. https://doi.org/10.1016/j.cep.2004.09.006
87. Gronald G, Derksen JJ (2011) Simulating turbulent swirling flow in a gas cyclone: a comparison of various modeling approaches. Powder Technol 205:160–171. https://doi.org/10.1016/j.powtec.2010.09.007
88. Parvaz F, Hosseini SH, Elsayed K, Ahmadi G (2020) Influence of the dipleg shape on the performance of gas cyclones. Sep Purif Technol 233:116000. https://doi.org/10.1016/j.seppur.2019.116000
89. Ganser GH (1993) A rational approach to drag prediction nonspherical particles. Powder Technol 77:143–152. https://doi.org/10.1016/0032-59109(93)80051-b
90. ESSS-Rocky (2015) Rocky-Dem Technical Manual. In: Models used in Rocky. ESSS, Rio de Janeiro, Brazil
91. Nogueira LW (2019) WORKSHOP 15 DEM-FEA one way coupling with ANSYS transient structural (workbench)
92. Almeida L (2019) Workshop 13—DEM-CFD one way coupling with ANSYS Fluent (workbench). In: Rocky Dem Part. Simulator-technical content. www.support.esss.co/hc/en-us/categories/115000554226-ROCKY. Accessed 27 Feb 2019
93. Almeida L (2019) Workshop 01 – transfer chute R4.2. In: Rocky Dem Part. Simulator-technical content. https://support.esss.co/hc/en-us/articles/115002452626-Workshop-01-Transfer-Chute. Accessed 21 Feb 2019
94. Almeida L (2019) Workshop 08 – Cone Crusher (Breakage) R4.2. In: Rocky Dem Part. Simulator-technical content. https://support.esss.co/hc/en-us/articles/115002453566-Workshop-08-Cone-Crusher-Breakage-. Accessed 21 Feb 2019
95. Walton OR, Braun RL (1986) Viscosity, granular-temperature, and stress calculations for shearing assemblies of inelastic, frictional disks. J Rheol (N Y N Y) 30:949–980. https://doi.org/10.1122/1.549893
96. Tsuji Y, Kawaguchi T, Tanaka T (1993) Discrete particle simulation of two-dimensional fluidized bed. Powder Technol 77:79–87. https://doi.org/10.1016/0032-5910(93)85010-7
97. Ma H, Zhao Y, Cheng Y (2019) CFD-DEM modeling of rod-like particles in a fl uidized bed with complex geometry. Powder Technol 344:673–683. https://doi.org/10.1016/j.powtec.2018.12.066
98. Zhou L, Zhang L, Bai L et al (2017) Experimental study and transient CFD/DEM simulation in a fluidized bed based on different. R Soc Chem Adv 7:12764–12774. https://doi.org/10.1039/C6RA28615A
99. Alobaid F, Ströhle J, Epple B (2013) Extended CFD/DEM model for the simulation of circulating fluidized bed. Adv Powder Technol 24:403–415. https://doi.org/10.1016/j.apt.2012.09.003
100. Razgaitis R, Guenther DA (1981) Separation efficiency of a cyclone separator with a turbulence-suppressing rotating-insert. J Eng Power 103:566–571. https://doi.org/10.1115/1.3230768

101. Baltrenas P, Pranskevicius M, Venslovas A (2015) Optimization of the new generation multi-channel cyclone cleaning efficiency. Energy Procedia 72:188–195. https://doi.org/10.1016/j.egypro.2015.06.027
102. Vaitiekūnas P, Petraitis E, Chlebnikovas A (2014) Cleaning efficiency researches in the cyclone with curved channels from different size particulate matter. In: The 9th international conference "environmental engineering", Vilnius, Lithuania
103. Ramlan R, Ngadiman Y, Omar SS (2015) Quantification of machine performance through overall equipment effectiveness. In: 2nd international symposium on Technology Management & Emerging Technologies: art proceeding (ISTMET). IEEE, Langkawi, Kedah, Malaysia, pp 429–433
104. Samat HA, Kamaruddin S, Azid IA (2012) Integration of overall equipment effectiveness (OEE) and reliability method for effectiveness. South Afr J Ind Eng 23:92–113. https://doi.org/10.7166/23-1-222
105. Coates W (1990) Jojoba harvesting equipment. Deciduous Fruit Nut P-83:91–93
106. Coates W, Lorenzen B (1990) Equipment for ground harvesting jojoba seed. Appl Eng Agric 6:125–130. https://doi.org/10.13031/2013.26358
107. Fonte CB Jr, Oliveira JAA, De Almeida LC (2015) Dem-Cfd coupling: mathematical modelling and case studies using Rocky-Dem® and Ansys Fluent®. In: Eleventh international conference on CFD in the minerals and process industries. CSIRO, Melbourne
108. Almeida L (2015) Workshop 9—CFD-DEM 1-way coupling. In: Rocky Dem Part. Simulator. lucilla@rocky-dem.com
109. Chu KW, Wang B, Yu AB, Vince A (2009) CFD–DEM modelling of multiphase flow in dense medium cyclones. PowderTechnology 193:235–247. https://doi.org/10.1016/j.powtec.2009.03.015
110. Cortés C, Gil A (2007) Modeling the gas and particle flow inside cyclone separators. Prog Energy Combust Sci 33:409–452. https://doi.org/10.1016/j.pecs.2007.02.001
111. Kozołub P, Klimanek A, Białecki RA, Adamczyk WP (2017) Numerical simulation of a dense solid particle flow inside a cyclone separator using the hybrid Euler–Lagrange approach. Particuology 31:170–180. https://doi.org/10.1016/j.partic.2016.09.003
112. Peng W, Hoffmann A, Boot PJA et al (2002) Flow pattern in reverse-flow centrifugal separators. Powder Technol 127:212–222. https://doi.org/10.1016/s0032-5910(02)00148-1
113. Li S, Yang H, Zhang H et al (2009) Measurements of solid concentration and particle velocity distributions near the wall of a cyclone. Chem Eng J 150:168–173. https://doi.org/10.1016/j.cej.2008.12.019
114. Wang L, Parnell CB, Shaw BW, Lacey RE (2006) A theoretical approach for predicting number of turns and cyclone pressure drop. Trans ASABE 49:491–503. https://doi.org/10.13031/2013.20404
115. Wang B, Xu DL, Chu KW, Yu AB (2006) Numerical study of gas-solid flow in a cyclone separator. Appl Math Model 30:1326–1342. https://doi.org/10.1016/j.apm.2006.03.011
116. Derksen JJ, Van Den Akker HEA, Sundaresan S (2008) Two-way coupled large-eddy simulations of the gas-solid flow in cyclone separators. AICHE J 54:872–885. https://doi.org/10.1002/aic.11418
117. Derksen JJ, Sundaresan S, van den Akker HEA (2006) Simulation of mass-loading effects in gas-solid cyclone separators. Powder Technol 163:59–68. https://doi.org/10.1016/B978-008044544-1/50089-3
118. Martignoni WP, Bernardo S, Quintani CL (2007) Evaluation of cyclone geometry and its influence on performance parameters by computational fluid dynamics (CFD). Braz J Chem Eng 24:83–94. https://doi.org/10.1590/S0104-66322007000100008
119. Elsayed K, Lacor C (2009) Investigation of the geometrical parameters effects on the performance and the flow-field of cyclone separators using mathematical models and large Eddy. In: 13th international conference on aerospace sciences & aviation technology, Cairo

120. Baker RV, Hughs SE (1999) Influence of air inlet and outlet design and trash exit size on 1d3d cyclone performance. Trans ASAE 42:17–21. https://doi.org/10.13031/2013.13193
121. Baker RV, Hughs SE, Gillum MN, Green JK (1996) Influence of TRASH exit design on cyclone performance. In: Beltwide Cotton Conference. National Cotton Council, Memphis TN, pp 1605–1609
122. Elsayed K, Lacor C (2010) The effect of vortex finder diameter on cyclone separator performance and flow field. In: Pereira JCF, Sequeira A (eds) V European conference on computational fluid dynamics. ECCOMAS CFD, Lisbon, pp 14–17
123. Elsayed K, Lacor C (2014) Analysis and optimisation of cyclone separators geometry using RANS and LES methodologies. Turbul Interact Notes Numer Fluid Mech Multidiscip Des 125:65–74. https://doi.org/10.1007/978-3-662-43489-5

Chapter 8
Challenges and Future Outlook

CFD–DEM coupling has emerged as a promising technique for studying the behavior of agricultural particulate matter within aerodynamic systems. However, despite significant progress, there remains a dearth of accurate and efficient models for CFD–DEM simulations of these complex systems. The current limitations of computational resources and the intricate shapes of agricultural grains have resulted in laboratory-scale simulations that fall short of capturing the true dynamics of real-world particulate systems in nature or industry. To overcome these challenges, researchers are actively pursuing advanced approaches for determining contact models that can account for the irregular shapes of agricultural grains. This promises to revolutionize the field by enabling high-resolution simulations that can accurately capture the translation and rotation of particles in 3D. By harnessing the power of CFD–DEM coupling and refining contact models, it can gain a deeper understanding of the behavior of agricultural particulate matter. This, in turn, can lead to the development of more efficient and sustainable agricultural practices grounded in empirical data and rigorous scientific inquiry.

Several models have been utilized in agricultural particle flow to describe particle shapes and determine particle contacts. However, despite the diverse range of available models, each has pros and cons regarding accuracy, flexibility, numerical stability, and efficiency. Consequently, there remain significant challenges in modeling agricultural particle flow, especially when attempting to capture the true shape of the particles. Fortunately, recent years have witnessed the advent of new, cutting-edge models such as the bounded particle model and software, which enable the simulation of irregular particles according to their true, real-world shape. These models have proven immensely useful in simulating various problems in the agricultural process. However, significant effort is still required to develop even more efficient, accurate, and trustworthy models capable of representing true particle shapes and detecting their contact forces. Indeed, developing such models is an urgent need and one of the most intriguing research areas to explore in the coming

L. Zhou et al., *Discrete Element Method for Multiphase Flows with Biogenic Particles*, https://doi.org/10.1007/978-3-031-67729-8_8

years. With continued effort and innovation, one can unlock new insights and strategies for improving the agricultural particle flow and enhancing the crop yields.

Investigating the behavior of particulate systems using the CFD–DEM is challenging due to the substantial computational cost involved in simulating these systems. With conventional PC hardware and workstations, the simulations are limited to a few particles, which hinders the realistic depiction of agricultural processes such as planting, harvesting, and post-harvesting. However, the advent of parallel computing, utilizing multi-core CPUs and Graphics Processing Units (GPUs), has shown immense potential in accelerating simulations. Despite recent advancements, modeling the behavior of agricultural particulate matter in dispersed systems is still an open area of research. Treating large ranges in grain shapes and sizes in big-sized cyclone separators poses a significant challenge, demanding an efficient approach to simulate and model these systems. Therefore, enhancing numerical techniques' validity, accuracy, speed, and versatility remains a critical aspect that needs improvement. To further improve CFD–DEM coupling, non-drag laws, thermal fluid-particle coupling, and mass exchange between fluid and particles should be implemented. However, the three-way and the four-way coupling approaches are considered more precise and accurate for representing real modeling in many systems; they consume considerable computational time. Thus, these approaches require high-end computational technology for calculating the interactions between the fluid and particle phases, especially for systems that contain many small particles, such as fine particles' flow in cyclones.

Furthermore, the future research should seek efficient methods for multi-phase and multi-physics problems, which are often complexly coupled together. However, the diverse characteristics of agricultural grains, such as size, shape, surface roughness, density, friction coefficients, composition, and other factors, pose a significant limitation to the use of CFD–DEM. Notably, the computational cost remains a significant constraint in applying CFD–DEM, particularly for smaller agricultural grains, requiring large computation time for simulation. Therefore, there is a need for innovative solutions that will significantly reduce computational costs while improving the accuracy and versatility of numerical techniques in modeling and simulating agricultural processes.

Index

L. Zhou et al., *Discrete Element Method for Multiphase Flows with Biogenic Particles*, https://doi.org/10.1007/978-3-031-67729-8

The manufacturer's authorised representative in the EU is Springer Nature Customer Service Centre GmbH, Europaplatz 3, 69115 Heidelberg, Germany. If you have any concerns regarding our products, please contact ProductSafety@springernature.com

Printed and bound by CPI Group (UK) Ltd, Croydon, CR0 4YY
01/10/2026
02225966-0019